INSTRUMENTS
AND
MEASUREMENTS

CHARLES M. GILMORE
DIRECTOR OF ENGINEERING
CMR TELEMETRY

McGraw-Hill Book Company

Gregg Division

New York
St. Louis
Dallas
San Francisco
Auckland
Bogotá
Düsseldorf
Johannesburg
London
Madrid

Mexico
Montreal
New Delhi
Panama
Paris
São Paulo
Singapore
Sydney
Tokyo
Toronto

Library of Congress Cataloging in Publication Data

Gilmore, Charles Minot, (date)
  Instruments and measurements.

  (Basic skills in electricity and electronics)
  Includes index.
  1. Electronic instruments. I. Title. II. Series.
TK7878.4.G55   621.3815'4   79-1435
ISBN 0-07-023297-0

*This book is dedicated to the engineering
staff of Heath Company who, with the able
assistance of Mike Rockwell and Tom Yeager,
have made significant contributions to the
art of low-cost instrumentation.*

Acknowledgments
Students, teachers, school administrators, and industrial trainers have contributed to the
development of the *Basic Skills in Electricity and Electronics* series. Classroom testing of
preliminary editions has been conducted at the following sites:

Burr D. Coe Vocational Technical High School (East Brunswick, New Jersey)
Chantilly Secondary School (Chantilly, Virginia)
Nashoba Valley Technical High School (Westford, Massachusetts)
Platt Regional Vocational Technical High School (Milford, Connecticut)
United States Steel Corporation: Edgar Thomson, Irvin Works (Dravosburg, Pennsylvania)

The publisher gratefully acknowledges the helpful comments and suggestions received
from these participants.

Instruments and Measurements

1 2 3 4 5 6 7 8 9 0   W C W C   7 8 6 5 4 3 2 1 0 9

The editors for this book were Gordon Rockmaker and Mark Haas. The designer was Tracy A.
Glasner. The art supervisor was George T. Resch. The production supervisor was Kathleen
Morrissey. Cover photography by Martin Bough/Studios Inc. It was set in Electra by
Progressive Typographers.
Printed and bound by Webcrafters, Inc.

Contents

Editor's Foreword

The Gregg/McGraw-Hill *Basic Skills in Electricity and Electronics* series has been designed to provide entry-level competencies in a wide range of occupations in the electrical and electronics fields. The series consists of instructional materials geared especially for the career-oriented student. Each major subject area covered in the series is supported by a textbook, an activities manual, and a teacher's manual. All the materials focus on the theory, applications, and experiences required for those just beginning their chosen vocations.

There are two basic considerations in the preparation of educational materials for such a series: the needs of the learner and the needs of the employer. The materials in the series have been designed to meet those needs. They are based on many years of experience in the classroom and with electricity and electronics. In addition, these books reflect the needs of industry and commerce as developed through questionnaires, surveys, interviews with employers, government occupational trend reports, and various field studies.

Further refinements both in pedagogy and technical content resulted from actual classroom experience with the materials. Preliminary editions of selected texts and manuals were field tested in schools and in-plant training programs throughout the country. The knowledge gained from this testing has enhanced the effectiveness and the validity of the materials.

Teachers will find the materials in each of the subject areas well coordinated and structured around a framework of modern objectives. Students will find the concepts clearly presented with many practical references and applications. In all, every effort has been made to prepare and refine the most effective learning tools possible.

The publisher and editor welcome comments from teachers and students using this book.

Charles A. Schuler
Project Editor

BASIC SKILLS IN ELECTRICITY AND ELECTRONICS

Charles A. Schuler, Project Editor

Books in this series

Introduction to Television Servicing by Wayne C. Brandenburg
Electricity: Principles and Applications by Richard J. Fowler
Instruments and Measurements by Charles M. Gilmore
Microprocessors by Charles M. Gilmore (*in preparation*)
Motors and Generators by Russell L. Heiserman and Jack D. Burson
Small Appliance Repair by Phyllis Palmore and Nevin E. André
Residential Wiring by Gordon Rockmaker
Electronics: Principles and Applications by Charles A. Schuler
Your Future in Electricity and Electronics by William A. Stanton
Digital Electronics by Roger L. Tokheim

Preface

The subject of *Instruments and Measurements* is electronic instruments. The instruments introduced here are the very basic instruments. They are basic to the electronics industry in two ways. First, these instruments are used by nearly everyone working with electronics. You will find these instruments used in TV repair and electrical/electronic appliance repair as well as in the design and service of complex aerospace telemetry systems. Second, these instruments are the basis for many of the more complex instruments which are used in more complex applications. For example, the oscilloscope, which is covered in the last two chapters of this book, is used as a fundamental part of spectrum analyzers, curve tracers, and network analyzers, to mention just a few applications.

This book looks at just a few basic instruments. It covers the basic indicator and the basic analog meter. The basic analog meter is built into the passive multirange multifunction instrument called the VOM. Both analog and digital readout meters are discussed as are the electronic counter and the oscilloscope. Few instruments in use today do not depend on an understanding of these basic parts.

The purpose of this book is to study how certain instruments are built. Learning how to use instruments is much easier once you know what the instrument is and how it works. Another important objective is to learn what the instrument can and cannot do.

The preparation of this textbook involved many people. My thanks go to the instrument manufacturers who generously supplied information on their products. My thanks also go to Rose Price who typed and retyped the many versions of the manuscript. Finally a special acknowledgment and appreciation goes to my wife, Polly, who worked with me throughout this entire project and who provided invaluable editing assistance and advice.

Charles M. Gilmore

Safety

Electric devices and circuits can be dangerous. Safe practices are necessary to prevent electric shock, fires, explosions, mechanical damage, and injuries resulting from the improper use of tools.

Perhaps the greatest hazard is electric shock. A current through the human body in excess of 10 milliamperes can paralyze the victim and make it impossible to let go of a "live" conductor. Ten milliamperes is a small amount of electrical flow: It is *ten one-thousandths* of an ampere. An ordinary flashlight uses more than 100 times that amount of current! If a shock victim is exposed to currents over 100 milliamperes, the shock is often *fatal*. This is still far less current than the flashlight uses.

A flashlight cell can deliver more than enough current to kill a human being. Yet it is safe to handle a flashlight cell because the resistance of human skin normally will be high enough to greatly limit the flow of electric current. Human skin usually has a resistance of several hundred thousand ohms. In low-voltage systems, a high resistance restricts current flow to very low values. Thus, there is little danger of an electric shock.

High voltage, on the other hand, can force enough current through the skin to produce a shock. The danger of harmful shock increases as the voltage increases. Those who work on very high-voltage circuits must use special equipment and procedures for protection.

When human skin is moist or cut, its resistance can drop to several hundred ohms. Much less voltage is then required to produce a shock. Potentials as low as 40 volts can produce a fatal shock if the skin is broken! Although most technicians and electrical workers refer to 40 volts as a *low voltage*, it does not necessarily mean *safe voltage*. You should,

therefore, be very cautious even when working with so-called low voltages.

Safety is an attitude; safety is knowledge. Safe workers are not fooled by terms such as *low voltage*. They do not assume protective devices are working. They do not assume a circuit is off even though the switch is in the OFF position. They know that the switch could be defective.

As your knowledge of electricity and electronics grows, you will learn many specific safety rules and practices. In the meantime:

1. Investigate before you act
2. Follow procedures
3. When in doubt, *do not act*: Ask your instructor

GENERAL SAFETY RULES FOR ELECTRICITY AND ELECTRONICS

Safe practices will protect you and those around you. Study the following general safety rules. Discuss them with others. Ask your instructor about any that you do not understand.

1. Do not work when you are tired or taking medicine that makes you drowsy.
2. Do not work in poor light.
3. Do not work in damp areas.
4. Use approved tools, equipment, and protective devices.
5. Do not work if you or your clothing are wet.
6. Remove all rings, bracelets, and similar metal items.
7. Never assume that a circuit is off. Check it with a device or piece of equipment that you are sure is operating properly.

8. Do not tamper with safety devices. *Never* defeat an interlocking switch. Verify that all interlocks operate properly.

9. Keep your tools and equipment in good condition. Use the correct tool for the job.

10. Verify that capacitors have discharged. Some capacitors may store a lethal charge for a long time.

11. Do not remove equipment grounds. Verify that all grounds are intact.

12. Do not use adaptors that defeat ground connections.

13. Use only an approved fire extinguisher. Water can conduct electric current and increase the hazards and damage. Carbon dioxide (CO_2) and certain halogenated extinguishers are preferred for most electrical fires. Foam types may also be used in some cases.

14. Follow directions when using solvents and other chemicals. They may explode, ignite, or damage electric circuits.

15. Certain electronic components affect the safe performance of the equipment. Always use the correct replacement parts.

16. Use protective clothing and safety glasses when handling high-vacuum devices such as television picture tubes.

17. Do not attempt to work on complex equipment or circuits before you are ready. There may be many hidden dangers.

18. Some of the best safety information for electric and electronic equipment is in the literature prepared by the manufacturer. Find it and use it!

Any of the above rules could be expanded. As your study progresses, you will learn many of the details concerning proper procedure. Learn them well, because they are the most important information available.

Remember, always practice safety; your life depends on it.

Simple Indicators

■ This chapter discusses two types of indicators, the neon-lamp indicator and the LED indicator. Indicators are vital in making electrical and electronic measurements. In this chapter you will learn the basic principles of both the neon lamp and the LED. You will also learn to draw diagrams for testing circuits with these indicators.

1-1 INTRODUCTION

Electrical and electronic measurements give answers to certain questions about a circuit. For example, a simple question asks, "Is there a voltage in the circuit?" A more complicated question asks, "Is the voltage ac or dc?" or "What is the polarity of the dc voltage?" These questions may be answered by using a simple indicator such as a neon lamp or a light-emitting diode (LED).

The neon-lamp indicator is used by electricians and others working on high-voltage circuits. Often these circuits are ac power lines. Special LED indicators are used to help design, test, and service logic circuits. In fact, the LED indicator is the fundamental building block of most logic probes.

1-2 THE NEON LAMP AS AN INDICATOR

The neon lamp is a simple but very good voltage indicator. Neon indicators are often found in electricians' tool kits. Most of the time the electrician wants to know if the circuit is "hot." That is, does the circuit have voltage or not. If the circuit is hot, the voltage is probably right. Therefore, the electrician does not need to measure the actual amount of voltage. A calibrated voltmeter is not necessary.

The neon lamp can tell you many facts about the voltage in a circuit. It can tell you:

1. If a voltage is greater than a given value
2. The polarity of dc voltage
3. Whether the voltage is ac or dc

Neon-Lamp Fundamentals

To understand how the neon lamp can do all this, we must review the fundamentals of the neon lamp. Figure 1-1 shows the V-I (voltage versus current) curves for a neon lamp. As you increase the voltage across the lamp, no current flows until the firing voltage is reached. For voltages below the firing voltage the neon gas has almost infinite resistance. When the firing voltage is reached, the neon gas will ionize.

The neon gas becomes ionized because of the strong voltage field. An arc is made in the ionized gas which carries a current. The arc

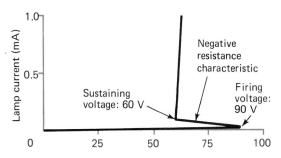

Fig. 1-1 The V-I (voltage versus current) curves for an NE-2 neon lamp. The voltage across the lamp is increased to 90 V, at which point the lamp fires. After the lamp fires, the current increases and the lamp voltage drops to 60 V.

resistance is much less than the resistance of nonionized gas. Neon lamps fit into a category known as *negative-resistance devices*.

When the lamp fires, the voltage across the lamp drops to the lowest value needed to keep the arc going. Something else also happens when the neon gas ionizes and arcs: The neon gas has the familiar red-orange neon glow.

Because the ionized neon gas resistance is low, current in the lamp must be limited to a safe level; otherwise the lamp would burn up. A common circuit used to limit neon-lamp current is shown in Fig. 1-2. The resistor in series with the lamp limits the lamp current to a safe value. For most common neon lamps this is 1 or 2 mA (milliamperes).

For example, suppose you wish to operate an NE-2 neon lamp with a 1-mA current from a 160-V (volt) source. You would do the following:

1. You would find the sustaining voltage for the NE-2 is 60 V by using the V-*I* curves in Fig. 1-1.
2. You know that once the lamp fires you want the lamp current to be 1 mA.
3. You know that the current is the same everywhere in a series circuit, so that the series resistor will also carry a 1-mA current.
4. Therefore, you calculate the voltage across the resistor as the supply voltage (160 V) less the lamp voltage (60 V). This is 100 V.
5. By Ohms law, $R = V/I$, the resistance, in ohms (Ω), is

$$R = \frac{100 \text{ V}}{0.001 \text{ A}} = 100,000 \text{ } \Omega$$

Before the lamp will fire, the voltage across the lamp must reach the 90-V firing voltage. You can see how the circuit conditions at

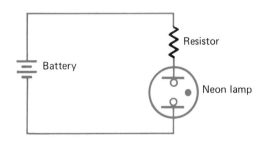

Fig. 1-2 A resistance placed in series with the neon lamp limits the current to a safe level. Greater currents would overheat the lamp, destroying it.

turn-on let this happen. Before the lamp fires, its resistance is very high. No current is drawn when the resistance of the lamp is very high. If no current flows in the lamp, no current flows in the series resistor. Therefore, there is no voltage drop across the resistor.

The voltage across the neon lamp tries to reach 160 V because this is the full power-supply voltage. But, when the voltage reaches 90 V, the lamp fires. When the lamp fires, 1 mA flows in the circuit. The current is kept at 1 mA because of the series resistor and the neon lamp's 60-V sustaining voltage.

We now have the basics of one function of the neon indicator. If the voltage across the lamp is less than 90 V, the lamp never fires. If the lamp fires, there must be more than 90 V in the circuit.

With just a 90-V supply, the lamp glows quite dimly. If you have more than 160 V, the lamp will glow very brightly. So we even have some indication of "how much" (the amplitude) voltage is in the circuit.

Self Test

1. You are building a neon-lamp indicator. You want the lamp to glow quite brightly, so you decide to use a 2-mA lamp current. The resistor in series with your lamp and your 180-V supply is
   A. 180,000 Ω
   B. 100,000 Ω
   C. 60,000 Ω
   D. 40,000 Ω

2. If you put two neon lamps in series, you double the required sustaining voltage. That is, you need 120 V across the two lamps to keep the arc going. You would expect that the firing voltage is
   A. 90 V
   B. 120 V
   C. 180 V
   D. 240 V

3. When the neon lamp fires, the voltage across the lamp drops to 60 V and the
   A. Neon gas glows
   B. Lamp resistance becomes high
   C. Current is limited to 1 mA
   D. Lamp will overheat unless the arc resistance is limited

4. Neon lamps are not useful in circuits where the voltage is less than
   A. 60 V   C. 120 V
   B. 90 V   D. 160 V

5. In an ac circuit the neon lamp fires when the voltage is greater than 90 V. An NE-2 is in series with a 47-kΩ resistor and is used in a 120-V rms ac circuit. The peak lamp current is
   A. 0.6 mA
   B. 1.0 mA
   C. 1.7 mA
   D. 3.3 mA

1-3 USING THE NEON LAMP TO TEST POLARITY

The neon lamp has another feature which makes it very useful. The neon lamp has two electrical elements. They are called the anode and the cathode. The *cathode* is the lamp element which is connected to the negative voltage. The neon gas glows only around the cathode. The *anode* is the element connected to the positive voltage.

Figure 1-3 shows the construction of the NE-2, which is a typical neon lamp. When the NE-2 is connected in a circuit, one of the elements is the anode and the other is the cathode. The NE-2 is now a dc polarity indicator.

For example, in Fig. 1-4 a neon lamp is connected across a 160-V dc supply. Looking at this figure you can see that one electrode is glowing. Therefore, the bottom connection is the negative terminal of the 160-V supply. If the leads of the neon lamp are reversed, the other electrode becomes the cathode. It will then be the element which glows. The neon lamp can therefore serve as a polarity indicator.

In Fig. 1-5 we can see that both neon-lamp elements glow. What does this tell us? If both

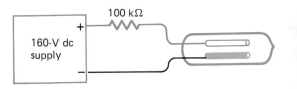
Fig. 1-4 The neon gas glows around the cathode (negative electrode) of the neon lamp.

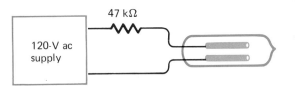
Fig. 1-5 Both neon-lamp electrodes glow. From this we can tell ac is connected to the lamp.

Cathode

Anode

Polarity indicator

Ac/dc indicator

elements are glowing, both are cathodes. But we know both lamp elements cannot be connected to the negative potential. Therefore, we must assume the lamp is connected to an ac supply. This is the reason both elements glow. When the lamp is connected to ac, one element is the cathode on the positive half cycle and the other is the cathode on the negative half cycle. At 60 Hz, it appears to the human eye that both elements are glowing continuously. The neon lamp is therefore also an ac/dc indicator.

This is shown in Fig. 1-6. Here you can see how one element glows for the positive half cycle and the other element glows for the negative half cycle.

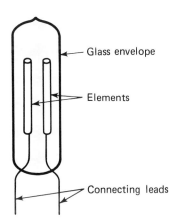
Fig. 1-3 The NE-2 neon lamp. The two metal elements (or electrodes) are sealed in a glass envelope or bulb. Before sealing, air is removed from the envelope and replaced with neon gas.

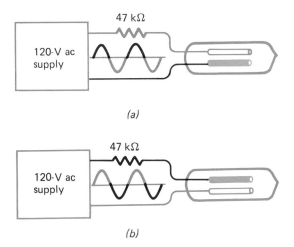
(a)

(b)

Fig. 1-6 The neon lamp connected to an ac supply. (*a*) You can see that the bottom electrode is the cathode in the first half of the cycle. (*b*) This is reversed in the next half of the cycle.

3

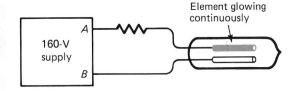

Fig. 1-7   Self test question 6.

LED (light-emitting diode)

Gallium arsenide

Gallium arsenide phosphide

Conduction knee

Self Test

6. The neon lamp in Fig. 1-7 is showing that
   A. Terminal A is the positive terminal of a dc supply
   B. Terminal B is one terminal of an ac supply
   C. Terminal A is the negative terminal of a dc supply
   D. Terminal B is the negative terminal of a dc supply

7. You have a neon-lamp tester constructed with three neon lamps in series with a resistor. All six elements are glowing. You know you must be connected to
   A. 100 V dc
   B. 200 V dc
   C. 120 V ac (rms)
   D. 240 V ac (rms)

8. You build a neon tester by putting two NE-2 lamps in series with a resistor. You connect this across a 190-V dc supply. Which elements in the lamps will glow? Draw a diagram of this circuit in operation and color the glowing elements red.

1-4 THE LED INDICATOR

The neon-lamp indicator must have a minimum of 90 V to fire the lamp. Many solid-state circuits operate on 5 to 20 V. The neon-lamp indicator is useless for checking these circuits because it takes too much voltage.

Fortunately, there is a low-voltage device which emits light when current is passed through it. This is the LED (light-emitting diode). As the name suggests, it has characteristics like the familiar silicon and germanium diodes.

LEDs are made using gallium arsenide (GAs) or gallium arsenide phosphide (GAsP). When a current passes through the LED, the diode junction emits light. Red is the most common color, but LEDs emitting green and yellow light are also available.

LED indicators are often used in solid-state circuits. Logic probes, for example, use the LED indicator because they work well with 5 V dc. Five volts dc is a common logic-circuit supply voltage.

The LED indicator can tell you many facts about the voltage in a circuit. It can tell you

1. If a voltage is greater than a given value
2. The polarity of a dc voltage
3. Whether the voltage is ac or dc

1-5 LED FUNDAMENTALS

Figure 1-8 shows the voltage-current curves of the LED, the silicon diode, and the germanium diode. The LED curves are much like those of the silicon diode. The voltage at which the LED conducts is higher than that of the silicon diode, but otherwise the curves are nearly identical. The voltage at which the diode conducts can be seen in Fig. 1-8. It is called the *conduction knee* or just the knee. It is the point where the slope of the V-I curve becomes much steeper. When the slope of the V-I curve becomes much steeper, the resistance has become much less.

In the LED, as in the neon lamp, current must be kept at a safe value after conduction starts. Once again, a series resistor is used as a current-limiting device. The major difference between the LED and the neon lamp is that the LED is not a negative-resistance device.

Once the knee on the V-I curve of the diode (see Fig. 1-8) is reached, the diode starts to emit light. Again, as with the neon lamp, the amount of light depends somewhat on the amount of current flowing through the LED. Most LEDs operate at currents of 5 to 20 mA.

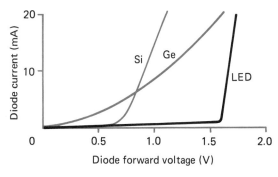

Fig. 1-8   The forward characteristic curves of the germanium (Ge), silicon (Si), and light-emitting diodes (LEDs). Note the LED has a higher junction potential and a lower on resistance.

For example, a TIL 209A LED takes 1.6 V to make a current of 20 mA flow. When this LED is used in a 5-V logic circuit, a series resistor limits the current to 20 mA. You calculate the value of the series resistor the same way for the LED as for the neon indicator. The current in the resistor is 20 mA. The voltage across the resistor is 5 V minus the LED voltage. This is 5.0 V minus 1.6 V, which is 3.4 V. By using Ohm's law, you can calculate the resistance value as

$$R = \frac{V}{I} = \frac{3.4 \text{ V}}{0.020 \text{ A}} = 170 \text{ } \Omega$$

A practical value is 180 Ω, because this is a standard value for a 5% resistor.

If the voltage across the LED is less than 1.6 V, there will be no current flow in the diode. Therefore, no light will be emitted. If the LED is dim, the low voltage is causing a low LED current. If the LED is very bright (or even getting hot), the voltage is high, making the LED draw too much current.

Self Test

9. An LED with a 1.6-V drop at 20 mA is to be used in a 15-V dc circuit. The standard value series resistance should be
   A. 220 Ω
   B. 470 Ω
   C. 680 Ω
   D. 1000 Ω

10. You have built a series circuit with four LEDs and a 470-Ω current-limiting resistor. This circuit gives a fairly bright indication when connected to the 5-V supply in a logic circuit. From this you would assume
    A. A current of 10.6 mA is flowing
    B. 5 V is probably a logical 1 in this circuit
    C. The 5-V supply has a bad regulator and is now 10 V or so
    D. 5 V is probably a logical 0 in this circuit

11. Most LEDs require between ____?____ and ____?____ mA of forward current.

12. An LED/resistor circuit is placed across a circuit with 5-V, 25% duty cycle pulses. During the pulse, the LED appears to glow ____?____ as brightly as if the 5 V were continuous.
    A. One half
    B. Just
    C. One quarter
    D. Three quarters

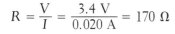

Fig. 1-9   Self test question 13.

13. If the circuit of Fig. 1-9 were used with a 220-Ω resistor, the LED current would be
    A. 100 mA
    B. 60 mA
    C. 20 mA
    D. 6 mA

1-6 USING THE LED TO TEST POLARITY

When the polarity of the voltage applied to the LED/resistor circuit is reversed, the LED does not emit light. However, the LED may draw current. Most LEDs have a reverse breakdown voltage of just a few volts. This is quite different from the familiar silicon diodes, which have reverse breakdown voltages in the tens or hundreds of volts.

After reverse breakdown is reached, the diode current must still be kept to a safe value. Otherwise the LED will be damaged from excessive power dissipation. In most cases, the series-limiting resistor does this.

Another reverse-voltage protection circuit is shown in Fig. 1-10. The silicon diode in series with the LED has a 75-V reverse breakdown voltage. The voltage required to forward-bias the circuit is now 1.6 V for the LED plus 0.7 V for the silicon diode. This gives a total of 2.3 V. The series resistor is

$$R = \frac{V}{I} = \frac{(15 - 2.3) \text{ V}}{0.020 \text{ A}}$$
$$= 635 \text{ } \Omega$$

The nearest standard-value 10% resistor is 620 Ω.

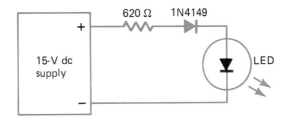

Fig. 1-10   Using a silicon diode in series with the LED to give a 75-V reverse breakdown.

Reverse breakdown voltage

Now, knowing the LED's reverse-polarity characteristics, you can test the polarity of a voltage source. If the LED emits light, the LED's anode is connected to a positive-voltage source. If the LED does not emit light, the LED's anode is connected to the negative terminal of the source. If the LED fails to light when you connect it to a circuit in one direction, but lights in the other direction, you know the polarity of the voltage source.

When you connect the LED tester to a circuit and it emits light when connected both ways, it is connected to an ac source. When the tester is connected to an ac source, the LED conducts for one half cycle like any other diode connected to an ac source. If you reverse the diode, the LED conducts for the other half cycle. For normal 60-Hz ac, the eye cannot tell which half cycle is causing the LED to light, or that the LED is actually blinking on and off 60 times per second. To the eye the LED appears to glow continuously. This operation is, therefore, just like the neon-lamp operation.

Often you cannot let the LED draw 5 mA or more from a circuit. Therefore, the LED is connected to the circuit with a buffer amplifier. This is shown in Fig. 1-11. A buffer amplifier prevents the LED from loading the

circuit. The amplifier can be as simple as just another gate. If the loading of the gate is too much, a special solid-state amplifier may be used. This is how many logic probes are built.

Many digital circuits have low-cost logic probes built into the product. These logic probes are usually nothing more than a resistor and an LED. Sometimes they are an LED/resistor circuit driven by a spare gate package.

Self Test

14. The LED in Fig. 1-12 is growing warm to the touch. It is not emitting light. This tells you that
    A. Terminal A is the positive terminal of a dc supply
    B. Terminal B is the positive terminal of a dc supply
    C. The terminals are those of an ac supply
    D. Terminal B is the negative terminal of a dc supply.

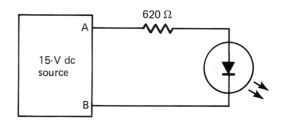

Fig. 1-12 Self test question 14.

15. The LED, like any other diode, has an anode and a cathode. The anode is made up of P-type material and the cathode is N-type material. When the LED is reverse-biased by a few volts, it
    A. Has a reverse current equal to its forward current
    B. Has no reverse current
    C. Has a slight reverse leakage current
    D. Emits light

16. Using an LED test circuit, you can tell that a voltage source is ac because the LED
    A. Pulses on and off
    B. Emits light when it is connected to the circuit either way
    C. Will be reverse-biased on negative half cycles
    D. Gets hot

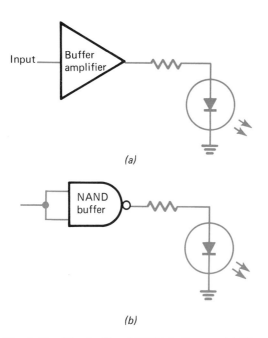

(a)

(b)

Fig. 1-11 The buffered LED indicator. (a) The LED is buffered with a simple amplifier. (b) The LED is buffered by an active output NAND gate.

6

1. The neon lamp makes a useful indicator to detect voltage in a circuit.

2. A neon lamp needs about 90 V across it to fire. That is, 90 V is needed to ionize the gas. Once the gas is ionized, the voltage across the neon lamp drops to 60 V.

3. Ionized neon gas has a low resistance. If the dc current is not limited, the lamp will draw too much current and burn up. To prevent this, a series resistance is added.

4. The neon lamp has two elements. The element which is connected to a negative potential is called the cathode. The element which is connected to the positive potential is called the anode.

5. When the lamp fires, the neon glow is around the cathode.

6. The LED (light-emitting diode) may be considered as a low-voltage neon lamp. That is, it glows after a certain voltage causes enough current to flow through it.

7. Most LEDs need about 1.6 V and 5 to 20 mA to emit light. A series resistor is also used in the LED circuit.

8. The amount of light from the LED depends on the LED current.

9. Reversing the polarity of the LED current means the LED will not emit light.

10. A series limiting resistor and a series silicon diode is used to protect the reverse-biased LED.

11. If the LED emits light, its anode is connected to the positive source.

12. If the LED emits light when it is connected either way to a source, it must be an ac source.

Chapter Review Questions

1-1. When the neon lamp ionizes, the resistance
(A) Increases (B) Decreases (C) Glows (D) Becomes infinite

1-2. If you have four neon lamps in series with a 220-kΩ resistor, you need at least _____?_____ V dc to draw a lamp current of 1 mA.
(A) 160 (B) 260 (C) 360 (D) 460

1-3. You can tell a neon lamp has fired
(A) Because of the orange glow (B) Because the cathode is connected to the negative source (C) Because the anode is connected to the positive source (D) Because the glow surrounds both elements

1-4. The resistor in series with the neon lamp is used
(A) To block high ac voltages (B) To limit the lamp current to a safe value (C) To prevent any current flow at low voltages (D) To divide high voltages

1-5. Typically a neon lamp will fire when the voltage across it reaches
(A) 60 V (B) 90 V (C) 120 V (D) 180 V

1-6. The anode of a neon lamp
(A) May glow when the lamp is connected to ac (B) Will be the only element which glows when it is connected to a positive dc source (C) Will be the only element which glows when it is connected to a negative dc source (D) Draws all the lamp current

1-7. When both elements in a neon lamp glow, you know
(A) The upper element is negative (B) The lower element is negative (C) The source is ac (D) The source is dc

1-8. Some LEDs are
(A) Silicon diodes which emit light when they are forward-biased
(B) Germanium diodes which emit light when they are reverse-biased (C) Gallium arsenide diodes which emit light when they are

reverse-biased (D) Gallium arsenide phosphide diodes which emit light when they are forward-biased

1-9. The most common LEDs emit
(A) Red light (B) Yellow light (C) Green light (D) Blue light

1-10. The forward voltage-drop across an LED is usually about
(A) 0.3 V (B) 0.7 V (C) 1.6 V (D) 3.2 V

1-11. The main difference between the LED and the neon lamp is
(A) The color of the emitted light (B) The lamp current (C) The voltage needed to make light (D) The size

1-12. Increasing the LED current will
(A) Decrease the voltage drop (B) Cause the color to change (C) Increase light output to some extent (D) Waste energy

1-13 A circuit has five LEDs and a 560-Ω resistor. A lamp current of 20 mA is caused by
(A) 10 V (B) 15 V (C) 20 V (D) 25 V

1-14. An LED can be used as a polarity indicator because
(A) Its cathode glows when it is connected to a negative source (B) It will be hot (C) It will glow when connected either way to an ac circuit (D) It will glow only when its anode is connected to the positive supply

1-15. A buffer amplifier is used with an LED indicator
(A) When the indicator is to be used in a high-voltage circuit (B) When the indicator is to be used in a logic circuit (C) When a gate package is not available (D) When the LED would draw too much current from the circuit

Answers to Self Tests

1. *C*	8. The one connected to	11. 5; 20
2. *C*	the − 190-V dc terminal	12. *B*
3. *A*	and the one connected	13. *C*
4. *B*	to the anode of the first	14. *B*
5. *D*	lamp.	15. *A*
6. *C*	9. *C*	16. *B*
7. *D*	10. *C*	

The Meter Movement

- This chapter discusses the analog meter movement, which is a pointer on a scale. Meters are essential in measuring the amount of voltage or current in a circuit.

   In this chapter you will learn how to read an analog meter. You will also learn the basic principles of three types of meter movements. You will be able to draw diagrams of meter circuits and discuss the major factors affecting meter accuracy.

2-1 INTRODUCTION

The neon lamp and the light-emitting diode as simple indicators answer the question, "Is there voltage in the circuit?" These simple indicators answer a simple question. However, they are not able to answer the question, "How much voltage (or current) is there in the circuit?" To answer this question we can use an analog meter (Fig. 2-1). The movement of this meter is an indicator which answers the question "How much?" by the position of a pointer on a scale or numbers on a display.

The analog meter movement uses a pointer on a scale and is one of the most common indicators used for electrical and electronic instruments. Digital meters are also used. The digital meter movement does not use a pointer on a scale but tells how much with numbers like a digital clock. The digital meter is covered in detail in a later chapter.

If we ask "How much?" we must follow it with "How accurate is the measurement?" We will look at three different ways meters are built and how accurately each one will answer the question "How much?"

2-2 READING THE ANALOG METER

Figure 2-2 shows an analog meter indicating different amounts of current. The pointer position on the meter scale does this. Note that the actual value read depends a little on the person reading the meter. You must decide what the reading is.

For example, both meters in Fig. 2-2 show a full-scale value of 10. However, on the first

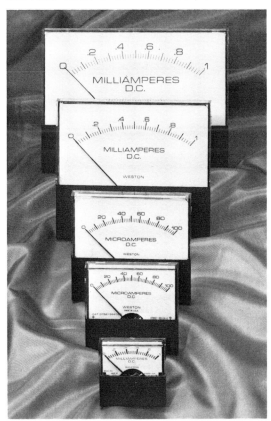

Fig. 2-1 Examples of analog meters. Analog meters are used on many different kinds of electronic instruments. (Courtesy of Weston Instruments Division of Sangamo Weston, Inc.)

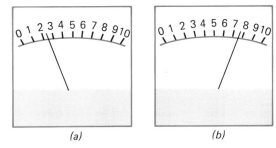

(a)   *(b)*

Fig. 2-2 Reading the analog meter. (*a*) The pointer lies just less than halfway between 2 and 3. The reading is 2.45. (*b*) The pointer rests just above the halfway mark between 7 and 8. The reading is 7.6.

Fig. 2-4

Fig. 2-5

Fig. 2-6

Fig. 2-7

meter, the pointer lies between 2 and 3. Reading the meter more closely, you see that the pointer lies at nearly 2½. An even closer reading of the meter's point leads to a final reading of 2.45.

Using the same method, the second meter reads about 7.6.

The meter pointer is not in direct contact with the meter scale. Therefore, its position on the meter scale depends on your viewing angle. Different readings are possible with changes in the viewing angle (Fig. 2-3).

This problem is known as *parallax*. It makes the reading of an analog meter subject to some error. Very-high-quality analog meters have a mirrored scale. This is so the person reading the meter can always be sure

to view the meter at the same angle. This is done by changing the viewing angle until the pointer's reflection in the mirror is blocked by the pointer itself. When this happens, it eliminates many of the parallax problems.

Self Test

1. Figures 2-4, 2-5, 2-6, and 2-7 show four different meter readings. Determine the reading for each figure.

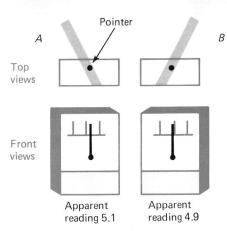

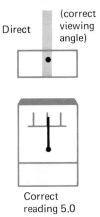

Fig. 2-3 Parallax. The eye at *A* sees the meter pointer over the 5.1 mark. The eye at *B* sees the pointer over the 4.9 mark. If the meter were observed directly, the pointer would show 5.0.

2. Parallax error happens because
   A. You always read the meter from the left-hand side
   B. You always read the meter from the right-hand side
   C. The pointer does not touch the meter scale, and therefore its position on the scale seems to change with viewing angle
   D. Analog meters have different scales

3. An analog meter displays quantity by
   A. A meter scale
   B. The position of a pointer on a meter scale
   C. Digits like a digital clock
   D. Measuring an electrical quantity

2-3 THE METER MOVEMENTS

Meters respond directly to the current passing through them. If you wish to measure a voltage instead of a current, the voltage must be converted to a current. Meters can measure current because they operate on the principles of magnetic repulsion. Meters are constructed in a number of different ways. They all depend on the magnetic forces of current in a coil, which repel (push away from) other magnetic forces.

There are three different types of meters. The lowest-cost and generally the most inaccurate is called the *iron-vane movement*. The most common is the moving-coil or the *d'Arsonval movement*. Both the iron-vane movement and the moving-coil movement use a single coil to generate a magnetic field. The *electrodynamometer movement* uses two coils. It is the most complex of the three meter movements and is normally used only for special applications.

The Iron-Vane Meter

Figure 2-8 is a diagram of an iron-vane meter. The iron-vane meter operates on the following principles:

1. An electric current in the coil magnetizes both the fixed and the moving vanes.
2. Because both of the vanes are magnetized by the same coil, they have the same magnetic poles.
3. One of the basic laws of magnetism states that like magnetic poles repel. This means they try to push apart.

The two like poles in the iron-vane meter movement are the two vanes. The moving

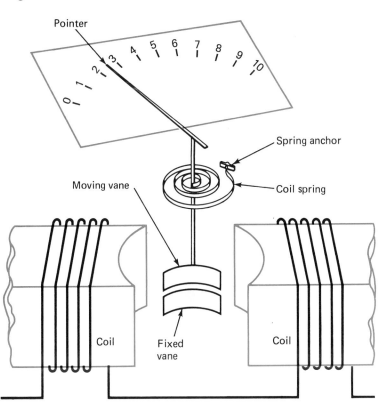

Fig. 2-8 The iron-vane meter movement. This exploded view shows the two vanes, the coils, the pointer, and the return spring.

vane is repelled by the fixed vane. The distance the moving vane is repelled depends on the strength of the magnetic field. The force generated by the magnetic field must work against the force of the coil spring. The coil spring is always trying to move the vane back to its original resting place. The spring returns the pointer to the zero position on the meter scale when there is no current in the meter coil.

What happens if the polarity of the current in the coil is reversed? The magnetic field is reversed. Therefore, the magnetic poles of the vane are reversed. However, even reversed, they are still alike and will repel each other. From this you can see that operation of the iron-vane meter movement does not depend on coil-current polarity.

Because operation does not depend on polarity, the iron-vane meter works on ac or dc. For this reason, the iron-vane meter is often used in ac applications where you cannot afford the greater cost of dc-only meters plus the additional cost of ac-to-dc conversion circuits.

Normal vanes move the meter pointer an equal distance on the scale for equal changes in coil current. That is, if the current doubles, the pointer moves twice as far. A special version of the iron-vane meter has shaped vanes. These shaped vanes give this meter movement a nonlinear response to current in

Coil spring

**Nonlinear
meter**

Fig. 2-9 A nonlinear meter scale. This meter is used to monitor line voltage which is nominally 120 V ac. The vanes of the meter are shaped so currents representing voltages of 0 to 90 V ac do not produce much meter movement, but currents representing 90 to 140 V move the pointer over two-thirds of the scale distance.

the coil. On a nonlinear meter, the amount of pointer movement for a change in coil current depends on the pointer's position on the scale.

An example of nonlinear meter movement is one which responds logarithmically. Often this is used to make a scale which shows decibels. Another example is a meter scale which is highly compressed at one end of the reading and quite expanded at the other. Figure 2-9 shows such a meter scale used for measuring line voltage. Notice that the expanded scale will make it easier to measure small changes in voltage near the correct line potential.

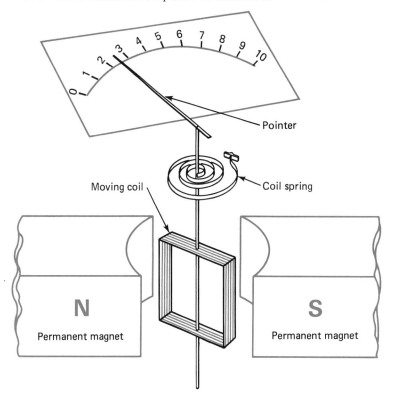

Fig. 2-10 The moving-coil or d'Arsonval meter movement. Note that the coil is attached to the pointer pivot.

The Moving-Coil Meter

Figure 2-10 shows the construction of a moving-coil meter. The basic operation of the moving-coil (d'Arsonval) meter is also simple. A coil attached to the pointer pivot is placed inside the field of a permanent magnet. The current in the coil generates a magnetic field. The magnetic field is repelled by the magnetic field of the permanent magnet. The current in the coil must be the right polarity. If it is not, it generates a magnetic field which attracts the field of the permanent magnet. This makes the coil and the pointer move in the wrong direction.

The more current there is in the coil, the stronger the coil's magnetic field. The stronger the coil's magnetic field, the farther the pointer is moved up the scale. Once again, the magnetic field is working against the force of the coil spring. The coil spring tries to return the pointer to the zero position.

Because the current in the coil must be the right polarity, the moving-coil meter responds only to direct current. If alternating current is applied to the moving-coil meter, the pointer moves up scale on positive half cycles and down scale on negative half cycles. This makes the meter vibrate. With high-frequency alternating current, the pointer will just stand still. If the moving-coil meter is to be used to measure alternating current, additional circuits must be added to convert ac to dc.

There are two different ways commonly used to build moving-coil meters. The older form of meter is called a *pivot-and-jewel*. The coil is mounted on a pivot. The pivot connects the coil, the pointer, and the spring. Each end of the pivot rests in jeweled bearings. The jeweled bearings let the pivot turn easily.

A newer form of the moving-coil meter is called the *taut-band meter*. The pointer and coil are mounted to a band of spring metal. The band of spring metal is stretched tightly between two supports. This taut-band meter is a simpler, more reliable, and much more rugged meter. It is rapidly replacing the older pivot-and-jewel meter construction.

The Dual-Coil Meter

The dual-coil meter, or electrodynamometer, is also an ac/dc meter (see Fig. 2-11). Notice that one coil is stationary. This is like the coil in the iron-vane meter. The other coil is moving. This is like the coil in the moving-coil meter. The magnetic fields of the two

d'Arsonval meter

Pivot-and-jewel meter

Taut-band meter

Electro-dynamometer

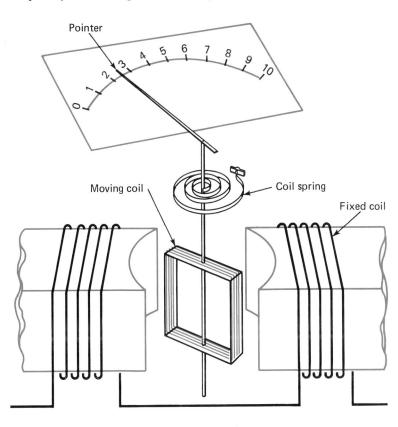

Fig. 2-11 The dual-coil or electrodynamometer movement. Note the two coils. Frequently the fixed coil has few turns and low resistance while the moving coil has many turns and high resistance.

Phase angle

Wattmeter

Rms value

Average current

Mean current

Accuracy

Meter rectifiers

Peak-responding

Average-responding

coils always repel each other. The repelling force is the force of the moving coil times the force of the fixed coil. Therefore, the meter pointer movement shows the product of the two coil currents.

There is one time when the repelling force is not the simple product of the two coil currents. This is when the phase angle of current in the stationary coil is different from the phase angle of current in the moving coil. In this case, the force, and therefore the meter pointer movement, depends on the currents and their relative phases.

The meter can be wired so the current in one coil is the current flowing to a load. You can also wire it so the current in the second coil is caused by the voltage across the load. This makes the meter deflection read the power in the load. In other words, for one coil the magnetic field is proportional to V. For the second coil the magnetic field is proportional to I. From the power formula, $P = VI$, the meter movement shows power. These meter connections are shown in Fig. 2-12. Such a meter is called a *wattmeter*.

Comparing the Meters

Both the iron-vane meter and the dual-coil meter may be used to measure the rms value of an ac or dc current. The current through a moving-iron-vane meter establishes the flux between the poles. A flux is induced in a moving vane and a flux of like polarity is induced in a fixed vane. The force of repulsion is due to the flux in one vane *times* the flux in

the other vane. Since the flux in each vane is proportional to the current, the pointer responds to the current times the current, or I^2. A pointer cannot respond instantaneously. Therefore, if the current is changing, the pointer will indicate the average (or mean) value of the I^2 value. Now, by calibrating the scale according to the square root of the mean I^2 value, we have an rms indicator. *Note:* Rms stands for the square root of the mean square value (root *m*ean square). The same effect takes place in the dual-coil meter when the flux is generated between two coils carrying the same current. The d'Arsonval movement uses a fixed or permanent flux in opposition to a current-produced flux. Therefore, it is simply a current indicator and not an I^2 indicator. It too cannot respond instantaneously and indicates the average value of the current.

The iron-vane meter is a low-cost, moderately accurate (5 to 10%) meter. The dual-coil meter is expensive, with typical accuracies from 0.2 to 3%. For these reasons most circuits have been developed to use the moving-coil meter. It is a moderate-cost, moderately accurate (1 to 3%) meter.

Ac Measurements with the Moving-Coil Meter

When the moving-coil meter is used to measure ac, the ac must first be converted to dc. This is normally done with meter rectifiers. Meter rectifiers are diodes selected for low forward voltage drop. Figure 2-13 shows a moving-coil meter used with a diode bridge. The direction of current flow through the meter is the same for both positive and negative half cycles of the ac waveform.

Meter circuits using the meter rectifier must have specially calibrated scales which read rms current. Depending on the type of meter rectifier circuit used, the meter may be either peak-responding or average-responding. Peak-responding meters have pointer deflection based on the peak value of the ac cycle. Average-responding meters have pointer deflection based on the average value of the positive (or negative) ac half cycle.

If a rectifier circuit is peak-responding, the scale will normally be calibrated to indicate 0.707 of the peak value. This is because for sinusoidal ac:

$$rms = 0.707 \times peak\ value$$

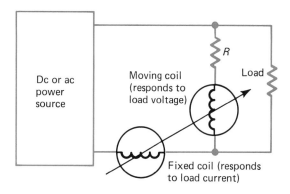

Fig. 2-12 The dual-coil meter is used to measure power. The fixed coil measures current to the load. The moving coil with the series resistance R carries a current proportional to load voltage. Meter deflection is proportional to the product of voltage and current, which is power. This meter may be used for ac or dc measurements.

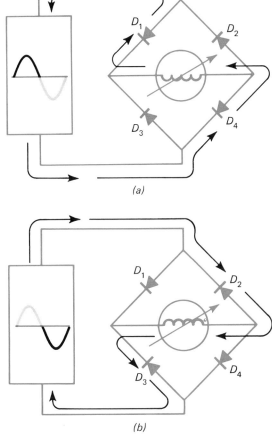

(a)

(b)

Fig. 2-13 Using the meter rectifier. On positive half cycles the diodes D_1 and D_4 direct current through the meter coil. On negative half cycles, the diodes D_3 and D_2 direct current through the coil. Meter movement is proportional to the average current in each half cycle.

If a rectifier circuit is average-responding, the scale will normally be calibrated to indicate 1.11 of the average value. This is because for sinusoidal ac:

$$\text{rms} = 1.11 \times \text{average value}$$

Thus we can obtain rms values for sinusoidal ac. For nonsinusoidal ac, *the meter will be incorrect*. Generally, a true rms meter is required to indicate the effective value for waveforms other than sinusoids.

For either sinusoidal or nonsinusoidal ac, the meter scale you want for most work is rms. This is because this is the way we most commonly think of ac. Any series resistance from the meter rectifier must be taken into account when determining the meter scales.

Meter movement of all three types are commonly housed in a *panel meter* (Fig. 2-14).

True rms meter

Panel meter

Fig. 2-14 The panel meter front and side views. Note the two center studs in the back of the meter. These are used to connect the wires. The studs in the corners are for mounting. (Courtesy of Weston Instruments Division of Sangamo Weston, Inc.)

This is the type found in electrical and electronic instrumentation. Note that the electrical terminals at the rear of the meter are designed for permanent connections.

Panel meters must be designed to be mounted in either steel or nonmagnetic materials. The magnetic properties of a steel panel change the accuracy of a meter designed to be mounted in a nonmagnetic panel such as aluminum.

Self Test

4. All common meters are based on the law of magnetism, which states that
   A. Like magnetic poles attract
   B. Opposite magnetic poles repel
   C. Electromagnetic poles repel
   D. Like magnetic poles repel

5. The d'Arsonval meter is constructed with
   A. Two coils, one fixed and one moving
   B. A fixed coil and a moving permanent magnet
   C. A fixed permanent magnet and a moving coil
   D. Two vanes, one fixed and one moving, and a single fixed coil

6. The electrodynamometer is constructed with
   A. Two coils, one fixed and one moving
   B. A fixed coil and a moving permanent magnet
   C. A fixed permanent magnet and a moving coil
   D. Two vanes, one fixed and one moving, and a single fixed coil

15

Coil resistance

7. The iron-vane meter is constructed with
   A. Two coils, one fixed and one moving
   B. A fixed coil and a moving permanent magnet
   C. A fixed permanent magnet and a moving coil
   D. Two vanes, one fixed and one moving, and a single fixed coil

8. The magnetic field from a coil is directly proportional to the strength of the current passing through it and the number of turns in the coil. Therefore, meters respond to
   A. Current
   B. Voltage
   C. Power
   D. Rms values

9. Taut-band construction is the ____?____ construction technique used for moving-coil meters.
   A. Older
   B. More sensitive
   C. Less reliable
   D. Most rugged

10. The current in a meter coil generates a magnetic field. This magnetic field is repelled by a second magnetic field, thus moving the meter's pointer. The force of these magnetic fields works against the force of ____?____ trying to return the pointer to the zero position.
    A. A permanent magnet
    B. An electromagnet
    C. A spring
    D. A pivot

11. The moving coil in a d'Arsonval meter responds to
    A. Average dc current
    B. Average ac current
    C. Rms dc current
    D. Rms ac current

12. A meter rectifier circuit is not usually
    A. Average-responding
    B. Peak-responding
    C. Rms-responding
    D. Used on moving-coil meters

13. If you need a meter movement which is low-cost, you select
    A. An iron-vane meter
    B. A moving-coil meter
    C. A moving-coil meter with a meter rectifier
    D. A dual-coil meter

14. If you need to make a power measurement, you select
    A. An iron-vane meter
    B. A moving-coil meter
    C. A moving-coil meter with a meter rectifier
    D. A dual-coil meter

15. If you want a medium-cost ac meter which is peak-responding but calibrated in rms, you choose
    A. An iron-vane meter
    B. A moving-coil meter
    C. A moving-coil meter with a meter rectifier
    D. A dual-coil meter

16. If you want a good medium-cost dc meter, you choose   .
    A. An iron-vane meter
    B. A moving-coil meter
    C. A moving-coil meter with a meter rectifier
    D. A dual-coil meter

2-4 MEASURING CURRENT

The basic discussions of meter principles show meter movements are current-measuring devices. A question which now must be answered is, "How much current forces the meter pointer on a normal meter to the full-scale position?" The answer to this question is, there is no "normal" meter. Different meter models need different amounts of current for a full-scale reading.

For example, one common meter requires 1 mA for a full-scale deflection. However, there are also more sensitive meters. They need less current for a full-scale deflection. There are less sensitive meters which require more current for a full-scale deflection.

Current does not flow through the meter coil completely unopposed. This is to say, the coil has resistance. Again, there is no standard value of meter resistance. Frequently, a 1-mA full-scale meter has a coil resistance of approximately 1000 Ω.

The current for full-scale deflection is very close to the same value from unit to unit of a particular meter model. However, the coil resistance may have a fairly wide tolerance. A 10% change or even a 20% change is common.

Ohm's law tells us there is a voltage-drop across the meter. For example, a 1-mA current passes through a meter with a coil resistance of 1000 Ω. The voltage-drop is

$$V = IR = 0.001\text{ A} \times 1000\text{ }\Omega = 1\text{ V}$$

If the meter resistance were exactly 1000 Ω, you could say the meter has a 1-V full-scale deflection. But you must remember the meter resistance may vary. Any variation in the resistance changes the full-scale voltage reading of the meter. The full-scale current reading of the meter is always 1 mA.

Using this meter, we can measure currents up to 1 mA. Currents less than 0.1 mA become difficult to read. This is because the meter deflection is less than one-tenth of full scale.

To measure currents greater than 1 mA with this meter, an additional circuit must be used. The concept is simple. A known percentage of the current to be measured is passed *through* the meter. A known percentage of the current to be measured is passed *around* the meter. One hundred percent of the current to be measured flows through the meter and the bypass circuits.

For example, let's use the 1000-Ω 1-mA meter from the previous example. We wish to measure currents up to 2 mA. At the maximum current (2 mA), we want to have 1 mA pass through the meter. Therefore, 1 mA must pass around the meter.

To pass current around the meter, a resistance is placed in parallel with the meter. This resistor is parallel with the meter is called a *shunt*. In the previous example, the shunt resistance is equal to the meter resistance. Therefore, one half the total current (50%) passes through the coil. The other half of the current (50%) passes through the shunt. This example is shown in Fig. 2-15.

Fortunately, we are not limited to shunts which only double the amount of current we can measure with the meter. The concept of the shunt may be extended as much as we need. For example, if the shunt resistance is one-third of the meter resistance, then one-fourth of the current passes through the meter and three-fourths of the current passes through the shunt. If this is applied to a 1-mA full-scale meter, it becomes a 4-mA full-scale meter. If a current of 2 mA is applied to this meter and shunt, the meter pointer moves to one half scale.

The shunt can carry many times the meter current. To make the shunt work, all you must know is the percentage of the current to

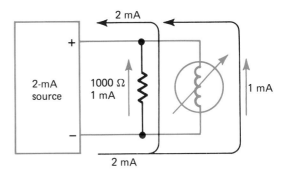

Shunt

Shunt resistance

Fig. 2-15 The shunt. A 1000-Ω resistor is in parallel with a 1-mA 1000-Ω meter. One half of the source current passes through the meter and one half through the shunt.

pass through the shunt and the percentage of current to pass through the meter. Remember, at all times 100% of the current must pass through the meter and the shunt system.

For example, a shunt carrying 999 times as much current as the meter makes an ammeter which measures 1000 times the full-scale meter movement value. Here we have 0.1% through the meter and 99.9% through the shunt. This equals 100% for the ammeter.

Let's look at this in more detail. The shunt must be equal to 1000 Ω divided by 999, that is, 1.001 Ω. If the shunt is placed across the 1000-Ω meter, 1 A is required in the combination meter and shunt for full-scale meter deflection. This example is illustrated in Fig. 2-16.

From these examples we may derive a formula to calculate shunt-resistance values. The shunt resistance is equal to the meter resistance divided by the portion of current to flow through the shunt. We can express this as

$$R_S = \frac{I_M \times R_M}{I_T - I_M}$$

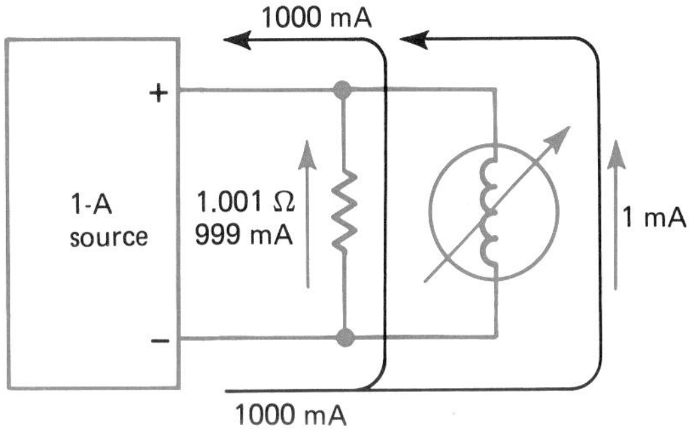

Fig. 2-16 The 1-A shunt. All but 1/1000 of the circuit current passes through the 1.001-Ω shunt resistor. A current of 1 mA (1/1000 of the circuit current) passes through the meter.

Current-measurement accuracy

Multiplier resistor

where R_S = the meter shunt resistance
R_M = the meter resistance
I_T = the total current (the desired ammeter full-scale value)
I_M = the meter movement's full-scale current

The formula is used in the following example: There is a need to measure currents of 0 to 800 mA. You have a 20-mA 200-Ω meter. The shunt must bypass 780 mA around the meter when the ammeter is measuring 800 mA. When this happens, 20 mA flows through the meter, giving a full-scale meter deflection. The full-scale pointer deflection indicates 800 mA in the ammeter. The shunt value may be calculated from our formula:

$$R_S = \frac{0.020 \text{ A} \times 200 \text{ }\Omega}{0.800 \text{ A} - 0.020 \text{ A}}$$

$$= \frac{4 \text{ V}}{0.780 \text{ A}} = 5.13 \text{ }\Omega$$

The 1 A and 800 mA examples show the large ratios of the meter resistance to the shunt resistance. The larger the ratio becomes, the more important it is that the shunt resistance remain very stable. If the ratio is large, for example, near 1000:1, a small change in shunt resistance has a major effect on the current-measurement accuracy. For this reason, the shunt must be carefully calculated knowing exact meter resistance. The value of the shunt resistance must also be stable with variations in temperature and over long periods of time.

Self Test

17. The value of the meter ____?____ is not always closely held.
    A. Coil resistance
    B. Full-scale current
    C. Scale size
    D. Pointer movement

18. A meter shunt is used when the meter full-scale current is ____?____ the maximum value of current to be measured.
    A. The same as
    B. Proportional to
    C. Less than
    D. Greater than

19. Given a 100-mA full-scale meter movement with a 10-Ω coil resistance, a shunt resistance of ____?____ Ω is required for a 1-A full-scale meter.
    A. 10
    B. 11.11
    C. 1.111
    D. 0.1

20. A meter has a 1-mA full-scale current. This meter is shunted by a 333.3-Ω resistor. The full-scale reading of the ammeter is ____?____ if the meter coil resistance is 1000 Ω. The full-scale reading is ____?____ if the meter coil resistance is 1200 Ω.

21. A 1-mA meter has a shunt which is $\frac{1}{9}$ of the meter coil resistance. To measure currents between 0.5 and 0.9 mA, how would you change the shunt resistance?

2-5 MEASURING VOLTAGE

As noted earlier, common meters measure current. To measure voltage, it must first be converted to an equivalent current and then applied to the meter. The way this is done is as simple as Ohm's law itself. To produce a known current from a known voltage, you use a known resistance.

For example, given a 1-mA 1000-Ω meter movement, a voltmeter with a full-scale deflection of 100 V is to be built. By Ohm's law,

$$R = \frac{V}{I} = \frac{100 \text{ V}}{0.001 \text{ A}} = 100,000 \text{ }\Omega$$

In other words, a 100,000-Ω resistor lets a 1-mA current flow when 100 V is applied to the resistor. However, the meter resistance is 1000 Ω. Therefore, we place a 99,000-Ω resistor in series with the meter to give a total resistance of 100,000 Ω. This additional resistance is called a *multiplier resistor*. A circuit using a multiplier resistor as described in this example is shown in Fig. 2-17.

The multiplier resistor plus the meter coil resistance convert a voltage to a proportional current. If 50 V is applied to this 100,000-Ω resistor/meter, 0.5 mA of current flows. The meter pointer deflects to only one half scale. You read this as 50 V.

As indicated earlier, the meter resistance is not always exact. Therefore, the multiplier-resistor value may have to be adjusted slightly if precise readings are required. Looking at the previous example, you can see that a fairly large change in the meter resistance will not cause any great difficulty.

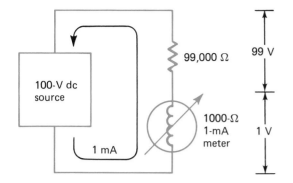

Fig. 2-17 The multiplier resistor. The sum of the 1000-Ω meter coil resistance plus the 99,000-Ω multiplier resistor gives a total of 100,000 Ω resistance. When 100 V is applied to the voltmeter, 1 mA flows and the meter deflects to full scale.

For example, if the meter resistance is 20% high (1200 Ω), the total resistance is now 100,200 Ω. Using Ohm's law to find the current generated by a 100-V source, we find

$$I = \frac{V}{R} = \frac{100 \text{ V}}{100,200 \text{ Ω}} = 0.000998 \text{ A}$$
$$= 0.998 \text{ mA}$$

This 20% change in meter resistance makes only a 0.2% change in the meter current. In this particular example, the meter multiplier resistor is very high when you compare it with the meter's coil resistance. Therefore, relatively large changes in meter coil resistance will not cause serious errors.

But when the meter multiplier resistor is much closer to the value of a meter coil resistance, changes in the meter's coil resistance become much more important. For example, if the 1-mA 1000-Ω meter is used to give a 2-V full-scale deflection, the total resistance must be

$$R = \frac{V}{I} = \frac{2 \text{ V}}{0.001 \text{ A}} = 2000 \text{ Ω}$$

We know the meter resistance is 1000 Ω. Therefore, the multiplier resistor must be 1000 Ω (1000 Ω + 1000 Ω = 2000 Ω). If the meter resistance now changes by 20% (1200Ω), the total resistance is now 2200 Ω. You can calculate the meter current changes using Ohm's law. That is,

$$I = \frac{V}{R} = \frac{2 \text{ V}}{2200 \text{ Ω}} = 0.000909 \text{ A}$$
$$= 0.909 \text{ mA}$$

The meter reads at $\frac{9}{10}$ of full scale for a 2-V

input. This is a 10% error. To maintain the greatest possible voltmeter accuracy, meter multiplier resistors are normally precision resistors.

Always remember, a current must be drawn from a circuit even to measure its voltage. The current is needed to move the meter. Wherever possible, a sensitive meter movement is used. This makes sure that the lowest possible current is drawn from the circuit being measured.

Self Test

22. The common analog meter measures current. A resistor in ___?___ with the meter is used to convert the desired voltage to the full-scale current value.
    A. Parallel
    B. Series
    C. Shunt
    D. Place

23. The value of the meter multiplier resistor is calculated so a maximum desired full-scale voltage produces a current through the meter equal to ___?___ times the full-scale meter current.
    A. 0.5
    B. 0.66
    C. 0.75
    D. 1.00

24. A 10-mA 2000-Ω meter movement is to be used to build a 200-V meter. A meter multiplier resistance of ___?___ is required.
    A. 2000 Ω
    B. 10,000 Ω
    C. 12,000 Ω
    D. 18,000 Ω

25. A 1-mA 1000-Ω meter uses a 99-kΩ multiplier to produce a 100-V full-scale meter. A 10% change in multiplier resistance produces a ___?___ change in the full-scale voltmeter reading.
    A. 0.1%
    B. 1%
    C. 10%
    D. 100%

26. A meter and multiplier resistor are used to construct a 100-V meter. A ___?___ full-scale meter draws the least current from the circuit under test.
    A. 100-μA
    B. 1-mA
    C. 10-mA
    D. 20-mA

2-6 METER-MOVEMENT ACCURACY

The commercial panel meter is not perfect. For example, the coil resistance is normally not closely held. Unfortunately, there are other sources of error that add to the inaccuracy of most meter readings.

Meter accuracy is normally specified as a percentage of a full-scale reading. Typical accuracies for today's moving-coil instruments are 2 to 3%. Most iron-vane instruments have 5 to 10% accuracies.

For example, a 0- to 10-mA meter is rated at 3% full-scale accuracy. This means a full-scale deflection will be a current of 10 mA ± 0.3 mA. At one-tenth of full-scale deflection, the current will be 1 mA ± 0.3 mA. This is a ± 30% error. Note that the accuracy is very poor for readings in the lower portion of the scale.

There are a number of other lesser specifications which also add to meter error. Some of these are:

1. *Repeatability:* Indicates how closely the meter returns to the same pointer position with the same coil current at different times. Bearing friction and ruggedness determine the meter repeatability.
2. *Temperature:* Most meters have a temperature-coefficient specification. It indicates variation in coil resistance with temperature. Depending on the circuit in which the meter is used, variations in coil resistance may or may not be a large part of the accuracy of the total reading.
3. *Position:* Most meters have a positional requirement. That is, the meter must be upright or lying flat to meet specifications. The bearings in most meters are not good enough to allow operation in any position.

Self Test

27. Given a 10-mA ± 2% meter, what is the range of actual current when the meter reads 4 mA?

28. Increased bearing friction should
    A. Increase meter accuracy
    B. Increase meter sensitivity
    C. Cause the meter to need a mirrored scale
    D. Decrease meter accuracy

29. Does a mirror-backed scale improve meter accuracy? Why?

30. Do some meters have accuracy specifications dependent on meter position? Why?

2-7 POWER MEASUREMENT

We have discussed use of the meter to measure current and voltage. We know that $P = VI$. This says that if current and voltage are known, the power applied to a circuit can be calculated.

In Fig. 2-18 a simple circuit containing two independent meters, a voltmeter and an ammeter, are used to measure the current through a load and the voltage across a load. Once the voltage and current are measured, you can use the power formula to calculate the power in a load.

For example, suppose the ammeter reads 100 mA and the voltmeter reads 100 V. The power in the circuit is

$$P = VI = 100 \text{ V} \times 0.1 \text{ A} = 10.0 \text{ W}$$

Note the voltmeter is placed directly across the load. The ammeter is placed between the voltmeter and the supply. The meter placement is chosen to eliminate the slight amount of voltage dropped by the ammeter itself. However, we must remember the ammeter is not only showing the current used by the load. It is also showing the current used by the voltmeter. The two possible solutions are diagramed in Fig. 2-19. The choice of which circuit you should use depends on the characteristics of the available instruments.

The simple voltmeter/ammeter power measurement becomes more difficult in ac cir-

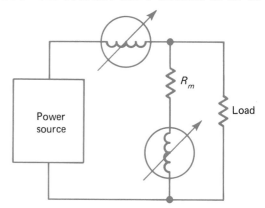

Fig. 2-18 A simple method to measure power. Power in the load is equal to the ammeter current (*I*) multiplied by the voltmeter reading (*V*).

Coil resistance

Meter accuracy

Percentage of full-scale

Repeatability

Temperature

Position

Two independent meters

Voltmeter/ammeter

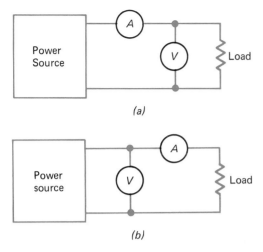

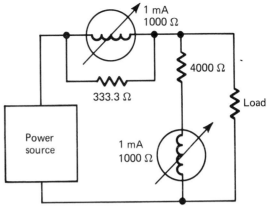

True power

Dual-coil
electro-
dynamometer

Fig. 2-20 Circuit for power measurement used in self test question 32. Both the ammeter and the voltmeter are reading full scale.

Fig. 2-19 Two power-measurement circuits. (a) In this circuit the power is in error by the power used in the voltmeter. (b) In this circuit the power measurement is in error by the power used in the ammeter. If the load power is high, the meter power is not significant.

cuits, where the load is not purely resistive. If the load is inductive or capacitive, the phase of a load voltage either leads or lags the phase of the load current. This makes the simple voltmeter/ammeter power-measurement technique no longer correct.

A true power measurement must take into account the relative phase of the voltage and current. The independent ammeter and voltmeter cannot do this. To get a true power measurement, the dual-coil electrodynamometer is used. The instrument, as was noted earlier, responds only to those portions of the two coil currents which are in phase.

Self Test

31. Using a simple voltmeter/ammeter circuit, a load with 50 V draws 200 mA. The load power is
   A. 250 W
   B. 4 W
   C. 10 W
   D. 2W

32. Given the voltmeter/ammeter circuit of Fig. 2-20, if the loading effect of the voltmeter is considered, the power in the load is ____?____ with full-scale meter reading on both meters.
   A. 1 W
   B. 0.5 W
   C. 0.2 W
   D. 0.015 W

33. A reactive ac load gives a 45° phase angle. A voltmeter/ammeter measurement reads 100 V at 1 A. The load power appears to be 100 W. The actual load power is ____?____.
   A. 100 W
   B. 70.7 W
   C. 10 W
   D. 1 W

34. A power measurement is to be made on an approximate 100-V dc, 1-A load. Two meters, one 0 to 100 mA at 10 Ω and one 0 to 100 μA at 100 Ω, are available. Draw a circuit to measure the load power exactly. Calculate the proper shunts and multiplier resistances. Indicate the meters to be used in each location. Why was each chosen?

Summary

1. The analog meter movement displays the amount of current being measured by changing position of a pointer on a scale. To read the analog meter, you first see approximately where the pointer is pointing, and then esti-

mate the pointer's actual position more and more closely.

2. The actual pointer position on the scale can appear to change because of your viewing angle. This is called parallax error.

3. There are three common meter movements: the iron-vane, the moving-coil, and the dual-coil, or electrodynamometer. All three depend on current in a coil creating a magnetic field. This magnetic field repels another magnetic field and forces the meter pointer to move up scale.

4. The iron-vane and dual-coil meters respond to the rms value of the ac or dc current in them.

5. The moving-coil meter measures dc only. Meter rectifiers are used with it to convert ac to dc to obtain an average value.

6. The current needed to produce a full-scale meter deflection is quite uniform from meter to meter. The coil resistance, however, may change quite a lot. Therefore, the meter deflection is specified in terms of current rather than voltage.

7. If the meter is to indicate currents greater than the movement full-scale capability, a shunt is used to bypass a known percentage of current around the meter.

8. Meter movements respond to current. Therefore, a voltage must be converted to a proportional current to be measured.

9. A meter multiplier resistor is put in series with the meter. This multiplier resistor plus the meter coil resistance converts the voltage to a proportional current.

10. Meter accuracy is given as a percentage of full scale. This tells you how close the meter reading is to the correct value, plus or minus a percentage of full-scale current.

11. Power is the product of the current and voltage in the load. A simple ammeter and voltmeter can be used to measure these quantities. The dual-coil meter corrects the phase-angle differences between the load current and voltage in ac loads.

Chapter Review

2-1. An analog meter movement answers the question "How much?"
(A) With numbers (B) By the position of a pointer on a scale (C) Within $\pm\frac{1}{2}\%$ (D) Within $\pm 1\%$

2-2. Parallax error
(A) Will cause a $\frac{1}{2}\%$ meter to become a $1\frac{1}{2}\%$ meter (B) Is caused by different reading angles (C) Is only found on high-quality meters (D) Is totally eliminated with a mirror-backed scale

2-3. Basically meter movements measure
(A) Voltage (B) Current (C) Power (D) Resistance

2-4. The ____?____ is not a common meter-movement construction.
(A) Moving-coil (B) Iron-vane (C) Electrostatic (D) Electrodynamometer

2-5. The iron-vane meter works because the magnetic fields of two vanes repel each other. These vanes are
(A) Magnetized by a permanent magnet (B) Magnetized by current in a coil (C) Shaped to give a nonlinear response (D) Dependent on the input current polarity

2-6. In all meter movements the pointer is returned to the zero position by
(A) A spring (B) A magnetic force (C) The operator (D) Gravity and friction

2-7. The iron-vane meter responds only to
(A) Positive dc currents (B) Negative dc currents (C) Ac currents (D) Ac and dc currents

2-8. The d'Arsonval meter uses
(A) Two vanes (B) Two coils (C) A moving coil and a fixed permanent magnet (D) A fixed coil and a moving permanent magnet

2-9. The moving-coil meter's pointer responds directly to
(A) Either ac or dc current (B) Dc current of either polarity (C) Dc current of one polarity (D) Ac current

2-10. The dual-coil meter
(A) Is an ac/dc meter (B) Is the lowest-cost meter (C) Is the only true rms meter (D) Is an average-reading meter

2-11. Meter rectifier circuits are used to
(A) Make dual-coil meters respond to dc (B) Make moving-coil meters respond to ac (C) Make iron-vane meters linear (D) Make moving-coil meters average-responding

2-12. Meters respond to current. To measure currents greater than the meter's full-scale value, you use
(A) A meter multiplier (B) A voltage divider (C) A 0.01-Ω resistor (D) A shunt

2-13. To double a meter's full-scale current you use a shunt whose value is _____?_____ the meter's coil resistance value.
(A) One-quarter (B) One-half (C) Equal to (D) Twice

2-14. A meter is usually calibrated to
(A) A known coil resistance (B) An exact full-scale voltage (C) An exact full-scale current (D) A known shunt resistance

2-15. If you want to build a 400-mA full-scale meter and you have a 10-mA 1000-Ω meter you will need a _____?_____ shunt.
(A) 1000-Ω (B) 256-Ω (C) 100-Ω (D) 25.6-Ω

2-16. To build a voltmeter you use a
(A) 1000-Ω resistor (B) Multiplier resistor (C) Shunt resistor (D) Parallel circuit

2-17. A voltmeter built with a meter movement and a resistor always
(A) Has an accuracy of $\frac{1}{2}$% (B) Draws little or no current from the circuit (C) Draws the full-scale meter current from the circuit (D) Draws enough current from the circuit to deflect the meter

2-18. Meter accuracy is usually specified as
(A) Friction (B) Linearity (C) A percentage of reading (D) A percentage of full scale

2-19. The simplest way to measure power is to
(A) Use the electrodynamometer (B) Use the voltmeter/ammeter (C) Use a phase-angle meter (D) Use a meter rectifier

Answers to Self Tests

1. Fig. 2-4: 4.5
   Fig. 2-5: 6.3
   Fig. 2-6: 9.4
   Fig. 2-7: 1.0
2. *C*
3. *B*
4. *D*
5. *C*
6. *A*
7. *D*
8. *A*
9. *D*
10. *C*
11. *A*
12. *C*
13. *A*

14. *D*
15. *C*
16. *B*
17. *A*
18. *D*
19. *C*
20. 4 mA, 4.6 mA
21. The shunt should be removed.
22. *B*
23. *D*
24. *D*
25. *A*
26. *A*
27. 3.8 to 4.2 mA
28. *D*

29. Yes, because it reduces parallax error.
30. Yes, because position changes the pivot friction.
31. *C*
32. *D*
33. *B*
34. The high-sensitivity meter was used as the voltmeter to avoid drawing extra current. The 100-mA 10-Ω meter was used so a practical shunt could be used.

The Volt-Ohm-Milliammeter (VOM)

■ This chapter discusses the volt-ohm-milliammeter (VOM). This instrument is sometimes called the volt-ohmmeter. The VOM is widely used in testing circuits.

In this chapter, you will learn the major features of the VOM as well as how to draw a schematic diagram of the common VOM circuits. You will also become familiar with the voltage, current, and resistance ranges of typical VOMs.

3-1 INTRODUCTION

In Chap. 2, we used the meter movement to answer the question, "How much voltage (or current) is there in a circuit?" We looked at the simple meter movement and saw that it responded to current. We found the amount of current that deflects the meter movement full-scale can be changed by connecting a shunt across the meter coil. You may make a meter movement measure voltage by using a multiplier resistor in series with a meter.

When you service electrical or electronic equipment, it is convenient to use one instrument. The VOM is a simple but very useful instrument for measuring ac and dc voltages, dc currents, and resistances. The VOM is a passive instrument, which means that it does not use vacuum-tube or transistor amplifiers to make it work. It consists of a meter, switches, resistors, and a battery.

When you know how an instrument works, you can make the best possible use of that instrument. Therefore, you will learn to diagram a VOM which can switch to a number of different voltage and current ranges.

3-2 A TYPICAL VOM

A typical VOM, which you might find in any electrician's tool box, a TV technician's tool kit, or an industrial electronics shop, is shown in Fig. 3-1. As you can see, the meter movement is housed in a rugged portable case which has a carrying handle. The top third of

the case front is covered by the meter movement and its scale. The rest of the case front has the range switch and the test-lead connections. This typical VOM will measure:

1. Dc voltage from less than 2.5 V to 5000 V
2. Ac voltage from less than 2.5 V to 5000 V
3. Dc current from less than 50 μA to 10 A
4. Resistances from a few tenths of an ohm to over 20 MΩ

To make these measurements, the VOM is a multifunction, multirange instrument. Multifunction indicates this instrument will measure more than one electrical quantity. By multirange we mean the VOM has more than one voltage range, more than one current range, and more than one resistance range. In fact, many different ranges are used. Most VOMs have enough ranges so that any reading in the instrument's complete range can be made in the upper part of the meter scale. As you remember from Chap. 2, readings at the upper end of the meter scale have less percentage error.

For example, if you wish to check a 24-V ac instrumentation transformer, you could use any range from 50 to 5000 V. The best range to use is the 50-V range, as it gives a midscale reading. If you wish to measure the 120-V ac power line, you should use the 250-V range.

By looking at the list of features for the VOM in Fig. 3-1, you can see the typical VOM does not include an ac current function. The ac current function is not normally

Fig. 3-1 The Weston Model 660 VOM. This typical VOM may be used for electrical and electronic servicing. (Courtesy of Weston Instruments Division of Sangamo Weston, Inc.)

found on VOMs. Most VOMs having low dc ranges (less than a few volts) do not have the same low-voltage ranges for the ac voltage function. The reasons this is so are explained in detail later in this chapter.

Some low-cost VOMs do not use switches to select the voltage, current, and resistance ranges. Instead, these instruments use many test-lead jacks to select the ranges. When using this kind of VOM, you plug in the test leads to a jack which selects the meter range. A switch is still used to select the functions such as dc voltage, ac voltage, or current. This does not change how the VOM works. It only changes how you operate it to make the measurement.

Self Test

1. The typical VOM ____?____ have the ranges necessary to work on automotive electrical systems.
   A. Does not
   B. Cannot
   C. Can
   D. Does

2. The ac ammeter function is ____?____ found on VOMs.

A. Typically
B. Never
C. Not typically
D. Always

3. The VOM could best be called a ____?____ instrument.
   A. Portable
   B. Fixed
   C. Bench
   D. Line-operated

4. The VOM is a passive instrument because
   A. It cannot be used to measure active devices
   B. It may or may not use a silicon diode in the meter circuit
   C. It is not constructed with active devices
   D. Under special circumstances a meter movement with rectifiers may be considered an active network

From page 24:
Multifunction

Multirange

VOM features

On this page:
Multirange
voltmeter

3-3 THE VOM VOLTMETER CIRCUIT

Figure 3-2 shows a simple circuit used to make a multirange voltmeter with a 50-microampere

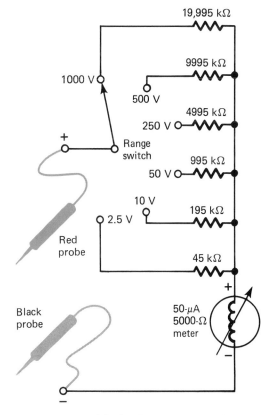

Fig. 3-2 A simplified VOM voltmeter schematic. This six-range voltmeter uses one 50-μA 5000-Ω meter movement, six parallel multiplier resistors, and one range switch. Some range switches add resistances in series so that lower value resistors may be used.

25

(µA) meter. When you set the range switch to a higher-voltage scale, you place more resistance in series with the meter movement. Ohm's law tells you that when you place more resistance in series with the meter, more voltage is required to produce the same current in the meter.

For example, Fig. 3-2 uses a 50-µA 5000-Ω meter movement. When 50 µA passes through the meter, the pointer is deflected to full scale. Therefore, if we wish to build a 50-V full-scale meter, the total circuit resistance (meter plus multiplier resistor) must cause a current of 50 µA when 50 V is connected to it. By Ohm's law,

$$R = \frac{V}{I} = \frac{50 \text{ V}}{50 \times 10^{-6} \text{ A}} = 1{,}000{,}000 \ \Omega$$

But the meter has a resistance of 5000 Ω; so the multiplier resistor must be

$$1{,}000{,}000 \ \Omega - 5000 \ \Omega = 995{,}000 \ \Omega$$

To make a 250-V full-scale meter, the resistance is

$$R = \frac{V}{I} = \frac{250 \text{ V}}{50 \times 10^{-6} \text{ A}} - 5000 \ \Omega$$
$$= 4{,}995{,}000 \ \Omega$$

a much higher resistance.

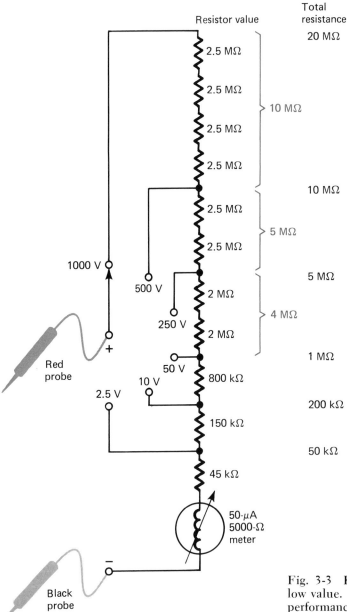

Fig. 3-3 Keeping the multiplier resistors at a low value. This circuit gives the same electrical performance as the circuit in Fig. 3-2, but no individual resistor is greater than 2.5 MΩ.

The number of resistors and the value of each resistor depend on the number of ranges and the exact range steps used by the VOM. In some VOMs the resistors may be connected in series. Some of the resistance may be made up of two or more resistors in series. Some of these combinations are shown in Fig. 3-3. This way of building a circuit makes the VOM easier to design. Very high values of precision resistors are difficult to get. The circuit in Fig. 3-3 keeps the maximum resistor size to 2.5 MΩ. This also keeps the maximum voltage across any resistor at 125 V or less. We can see this by using Ohm's law:

$$V = IR = 50 \times 10^{-6} \text{ A} \times 2.5 \times 10^{6} \text{ Ω}$$
$$= 125 \text{ V}$$

Frequently a precision resistor will change its value slightly when high voltages (greater than a few hundred volts) are applied. If this happens to the multiplier resistor, the accuracy of the VOM changes.

Self Test

5. A VOM has 1-, 3-, 10-, 30-, 100-, 300-, and 1000-V scales. Assume a 100-μA 1000-Ω meter movement is used. What are the resistances required to make the meter shown in Fig. 3-4?

6. What is the maximum voltage across any resistor in Fig. 3-4?

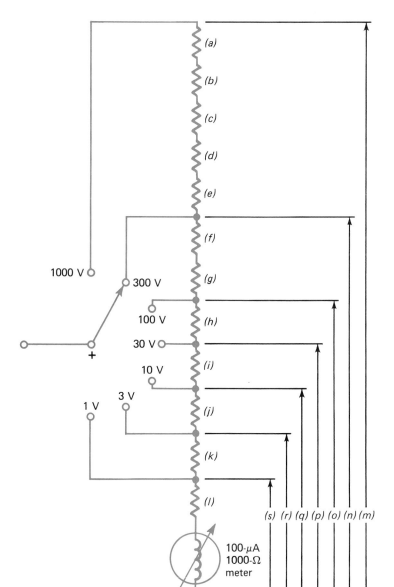

Fig. 3-4 Self test question 5.

27

Multirange
ammeter

7. What would happen to the resistance values if a 200-μA 1000-Ω meter was used?

3-4 THE VOM AMMETER

The simplified schematic in Fig. 3-5 shows one possible way to switch a multirange ammeter. As you switch in more sensitive current ranges, you switch in more shunt resistance. When you switch in more shunt resistance, more of the total current passes through the meter. This makes the meter more sensitive.

For example, in Fig. 3-5, a 50-μA 5000-Ω meter movement is used to make a multirange ammeter. When 50 μA passes through the meter, its pointer deflects the full scale. Therefore, if we are to build a 500-mA full-scale ammeter, 50 μA must pass through the meter movement and 500 mA minus 50 μA (499.95 mA) must pass through the shunt. By Ohm's law, we know the voltage-drop across the meter when it is carrying 50 μA, that is,

$$V = IR = 50 \times 10^{-6}\,A \times 5000\ \Omega = 0.25\ V$$

The voltage-drop across the shunt must also be 0.25 V. For a 500-mA range, the shunt is

$$R = \frac{V}{I} = \frac{0.25V}{500 \times 10^{-3}\,A - 50 \times 10^{-6}\,A}$$
$$= 0.50005\ \Omega$$

but for a 100-mA meter the shunt is

$$R = \frac{V}{I} = \frac{0.25V}{100 \times 10^{-3}\,A - 50 \times 10^{-6}\,A}$$
$$= 2.501\ \Omega$$

Of course, the most sensitive position is the one which completely eliminates the shunt.

This may or may not be done. The current range without shunts is very sensitive (between 50 and 100 μA) and can be easily burned out. The range switch must be designed with care. Good VOMs use a make-before-break range switch for the ammeter shunts. If this is not used, the full current can be applied to the meter movement between the range positions.

Note: You should never count on a make-before-break contact in the VOM's ammeter. Disconnect the load before switching ranges!

Self Test

8. Again using the 100-μA 1000-Ω meter to build a VOM, determine the required shunt resistances in Fig. 3-6. This VOM has 1-, 10-, 100-, and 1000-mA (1-A) ranges.

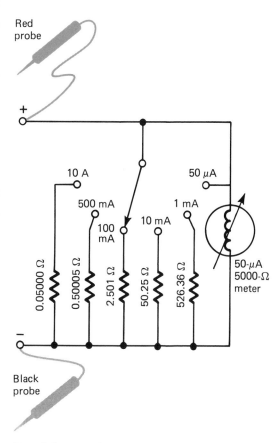

Fig. 3-5 A multirange ammeter. This circuit is typical of those found in many VOMs. The meter is a 50-μA full-scale, 5000-Ω movement.

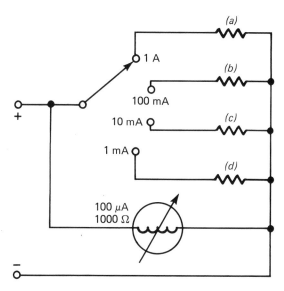

Fig. 3-6 Self test question 8.

9. What is the maximum current through any shunt resistor in question 8? Through the meter?

10. Why is a make-before-break switch used to select the ammeter shunts?

3-5 THE VOM AC VOLTMETER

You may make a multirange ac voltmeter by using the circuit shown in Fig. 3-7. Once again, a number of multiplier resistors are used. You can see the simple meter is replaced by a meter/meter-rectifier combination. The meter rectifier converts the meter movement into an ac responding meter. The meter rectifier is usually a germanium diode or a special-purpose meter-rectifier diode.

Before the meter-rectifier diodes will pass any current, a few tenths of a volt must be applied across them. In order to generate sufficient current to deflect the meter to full scale, 1.5 V or more must be applied to the meter/meter-rectifier combination. Frequently, the lowest ac voltage range has an individually calibrated multiplier. The meter scale may even have separate markings. This is done because the normal meter rectifier is not linear in the sub-1.5-V region.

The ac meter circuits frequently result in an instrument whose resistance is much lower than the high resistances found with a dc meter.

For example, assume your meter rectifiers are sufficiently linear to use a scale 0 to 2.5 V ac. We find the diodes must have 200 μA full scale to produce this linearity. By Ohm's law we know the total resistance on the 2.5-V range is

$$R = \frac{V}{I} = \frac{2.5V}{200 \times 10^{-6} \text{ A}} = 12,500 \text{ } \Omega$$

We know the meter used in our VOM is a 50-μA 5000-Ω meter. Therefore, it must be shunted to read full scale with 200 μA. The shunt is

$$R = \frac{V}{I} = \frac{0.25V}{200 \times 10^{-6}A - 50 \times 10^{-6}A}$$
$$= 1666 \text{ } \Omega$$

But we must also use a meter multiplier in series with this shunted meter to give the meter/shunt full-scale sensitivity of 2.5 V.

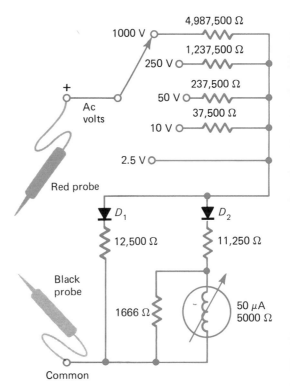

Fig. 3-7   A multirange ac voltmeter. Diodes D_1 and D_2 are the meter rectifiers which change the ac into dc for the 50-μA 5000-Ω meter movement.

Multirange ac voltmeter

Meter/meter-rectifier combination

Low voltage scales

The parallel resistances of the meter (5000 Ω) and the shunt (1666 Ω) are

$$R = \frac{R_1 \times R_2}{R_1 + R_2} = \frac{5000 \times 1666}{5000 + 1666} = 1250 \text{ } \Omega$$

A total resistance of 12,500 Ω is needed; so a multiplier resistance of

$$12,500 \text{ } \Omega - 1250 \text{ } \Omega = 11,250 \text{ } \Omega$$

is required. Of course, we must include the diode resistance in the 11,250 Ω.

Our meter movement receives the positive output from a half-wave rectifier. The negative half wave passes through D_2 and the 12,500-Ω resistor. The 12,500-Ω resistor and D_2 keep the loading effect uniform from the positive half wave to the negative half wave. Ideally, this resistor may be somewhat less than 12,500 Ω to allow for the series resistance that is actually in D_2.

The meter multiplier resistors may now be calculated. We will assume a 200-μA full-scale meter with a 12,500-Ω resistance. For example, the 10-V range requires that the

29

meter multiplier resistor drop 7.5 V at 200 μA. The multiplier is therefore

$$R = \frac{V}{I} = \frac{7.5V}{200 \times 10^{-6} \text{ A}} = 37,500 \ \Omega$$

You can see the multiplier resistors for the ac ranges are much smaller than those for the same dc ranges. This is because the ac meter really uses a 200-μA 12,500-Ω meter instead of the 50-μA 5000-Ω meter used by the dc voltmeter.

You can now see why an ac ammeter is not practical. A 50-μA 5000-Ω dc ammeter causes only a 250-mV drop when it is in the circuit. However, the ac ammeter causes a 2.5 V drop in the circuit.

Self Test

11. A VOM has 3-, 10-, 30-, 100-, 300-, and 1000-V ac ranges. Assume the meter/meter-rectifier combination gives a sensitivity of 200 μA and a series resistance of 15,000 Ω. What are the resistances required to complete the multirange meter shown in Fig. 3-8?

12. What is the maximum voltage across any resistor in Fig. 3-8?

13. If the meter diodes D_1 and D_2 are not ideal diodes, what resistance values in Fig. 3-8 change?

3-6 THE VOM OHMMETER CIRCUIT

Figure 3-9 is a schematic of a typical VOM ohmmeter circuit. As you can see, the ohmmeter consists of a battery, a switch-selected known resistance, and a variable resistor. These are all in series with R_{unknown}. The VOM meter movement has a single multiplier resistor which converts it into a 1.0-V full-scale voltmeter. A voltmeter measures the voltage developed across the switch-selected known resistance. The operation of this ohmmeter is really quite simple. To calibrate the meter, we short the test leads

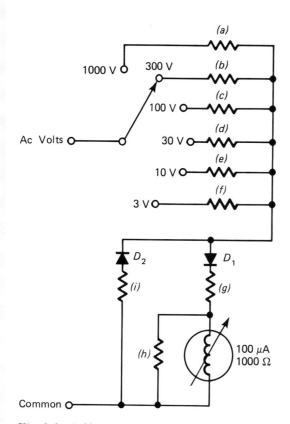

Fig. 3-8   Self test question 11.

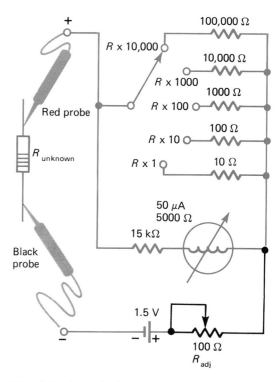

Fig. 3-9   A typical ohmmeter circuit used in VOMs. The variable resistor R_{adj} is used to compensate for anything which will change the current flowing in the ohmmeter circuit. The switched resistor $R \times 1$ through $R \times 10,000$ selects the meter's resistance ranges.

together. This makes $R_{unknown}$ 0 Ω. The current generated in the known resistor can be adjusted by the variable resistor R_{adj}. To calibrate the circuit, you set R_{adj} to cause a full-scale meter reading. This full-scale point is marked "0 Ω" on the meter scale.

If you put a resistor whose value is equal to the resistance in series with the meter between the test leads, you double the circuit resistance. If you double the circuit resistance, you divide the circuit current by 2.

If the current in the 10-Ω resistor is divided in half, then by Ohm's law we know the voltage across the 10-Ω resistor must fall to one-half the value which we found when $R_{unknown}$ was 0 Ω. The meter, which was at full scale, now reads one-half scale. If the resistance was 10 Ω, the center point on the scale can now be marked 10 Ω. Of course, this center point is now actually reading one-half the full-scale voltage. We have now calibrated the resistance scale of the meter in two places. Zero ohms is at the full-scale point. The center scale reads 10 Ω. As you can see, the ohmmeter scale is read opposite from all other meter scales. That is, 0 Ω is on the right, with higher values going left.

You may calibrate other points of the meter resistance scale in the same way. For example, use a value for $R_{unknown}$ which is nine times greater than the ohmmeter resistance. The total resistance will be $1R + 9R = 10R$. The total current will now be one-tenth of the full-scale current. Therefore, this point on the meter scale is marked with a value nine times greater than the center-scale value.

A typical ohmmeter is shown in Fig. 3-10. This simple circuit allows you to measure resistances over a 10 to 1 range. For example, this scale will easily show resistance of 10 Ω

(center scale) and resistance of 100 Ω (far left-hand scale).

When you want to measure greater resistances, you change the series resistance. For example, suppose the resistance is ten times greater than the original value. This new range is labeled the $R \times 10$ range. The original scale was marked the $R \times 1$ range. The original center-point scale was calibrated at 10 Ω. But with 100 Ω in series with the meter, the center-scale point is now 100, not 10. All other points in the ohmmeter scale are ten times greater than the $R \times 1$ markings. Of course, this resistance may also be made greater by a factor of 100, 1000, or even 10,000. This leads to resistance ranges which are labeled $R \times 1$, $R \times 10$, $R \times 100$, $R \times 1000$, and $R \times 10,000$. Looking at the ohmmeter schematic, you will see that a small variable resistor is added in series with the resistance. This variable resistor is added because the total series resistance on the low-resistance ranges changes. These changes are because of battery connections, test-lead connections, and internal VOM wiring. The variable resistor lets you correct for any of these day-to-day variations. It is used each time you change ohmmeter ranges, when you change the ohmmeter battery, when you turn on the ohmmeter, and each time the zero point changes.

Any time the VOM ohmmeter function is selected, the battery is in the circuit. When low-resistance scales are used, quite a bit of current is drawn from the ohmmeter battery. For example, when the test leads are shorted on a 10-Ω center-scale ohmmeter, the current drawn from the 1.5-V dry cell is

$$I = \frac{V}{R} = \frac{1.5V}{10 \; \Omega} = 150 \; mA$$

If the meter leads are left shorted too long, the battery will soon discharge.

When the VOM ohmmeter is used, a current is passed through the circuit being tested. Normally VOMs are wired so the common test lead is negative and the other lead is positive. Some VOMs reverse the polarity of the ohmmeter test current. You should check the VOM front panel for polarity markings. If it is not marked on the front panel, check the instruction manual. If necessary, check the polarity with another VOM.

R_{adj}

0 Ω position

10 Ω position

$R \times 10$ range

Ohmmeter polarity

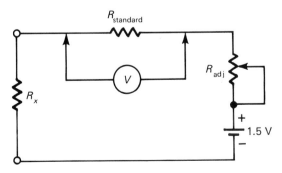

Fig. 3-10   The basic ohmmeter circuit.

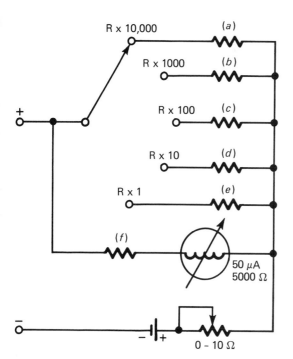

Fig. 3-11   Self test question 14.

14. The circuit in Fig. 3-11 uses a 50-μA 5000-Ω-full-scale meter. Fill in values a to f to create a 20-Ω center-scale ohmmeter.

15. What is the value of the $R \times 100$ resistor?

16. The ohmmeter scale is a ____?____ scale.
    A. Linear
    B. Proportional
    C. Inversely proportional
    D. Nonlinear

17. The variable resistor used in the VOM ohmmeter circuit does not compensate for
    A. Changes in battery voltage
    B. Changes in the ohmmeter internal circuit resistances
    C. Improperly marked resistors
    D. Small resistances found in the test leads

Summary

1. The VOM is a multipurpose portable instrument. It is used to measure ac and dc voltage, dc current, and resistance. The VOM is a passive instrument using only a meter and its switched shunts, multipliers, and ohmmeter circuit.

2. The voltmeter circuit in a VOM is made up of a high-sensitivity meter, a range switch, and a number of multiplier resistors. Each range may have its own multiplier resistor, or it may depend on the value of earlier multiplier resistors.

3. The VOM ammeter circuit allows you to switch different values of shunts in parallel with the VOM's meter movement. The higher the current to be measured, the lower the shunt resistance.

4. The ac voltmeter circuit in a VOM is built like the dc voltmeter. Added to the dc meter movement is a meter rectifier circuit, to make it respond to ac currents.

5. The ac meter does not have as much sensitivity as a dc meter.

6. The VOM ohmmeter circuit is used to measure resistance. Resistance is measured by comparing the unknown resistance with a known resistance.

Chapter Review Questions

3-1. A VOM is ____?____ instrument.
   (A) An active (B) A passive (C) A resistive (D) A capacitive

3-2. You could not use a common VOM to directly measure
   (A) The voltage at an ac household outlet (B) The current used by an electric toaster (C) The car battery's voltage (D) The current drawn by a digital integrated circuit (E) The continuity of a 60-W 120-V lamp

3-3. The VOM dc voltmeter is built using
   (A) Ammeter rectifiers (B) Multiplier resistors and a meter rectifier (C) Switched multiplier resistors (D) Switched shunt resistors

3.4. The highest-sensitivity VOM voltmeter is built using
(A) A high-sensitivity high-resistance meter (B) A high-sensitivity low-resistance meter (C) A low-sensitivity high-resistance meter (D) A low-sensitivity low-resistance meter

3-5. The VOM's ammeter uses a number of switched shunts. Using the shunts ___?___ the VOM's current sensitivity.
(A) Increases (B) Decreases (C) Does not change (D) Changes

3-6. If one ammeter range switch position does not use any shunt, the VOM's current sensitivity will be ___?___ the basic meter movement's sensitivity.
(A) Greater than (B) Less than (C) The same as (D) Unchanged by

3-7. The purpose of a meter rectifier in a VOM is to
(A) Convert ac into dc (B) Convert rms into peak-to-peak (C) Convert rms into average (D) Provide low-value ac voltage readings

3-8. The VOM ac voltage scales may be nonlinear on
(A) Ac signals (B) Dc signals (C) The highest ranges (D) The lowest ranges

3-9. A VOM ohmmeter has a nonlinear scale because the circuit ___?___ is changed by an unknown resistor.
(A) Current (B) Voltage (C) Linearity (D) Scale

3-10. If the unknown resistor is four times as large as the ohmmeter's calibrating resistor, the meter pointer will be at ___?___ of full scale. This point on the meter scale is marked with a resistance which is four times the center-scale value.
(A) One-half (B) One-third (C) One-fourth (D) One-fifth

3-11. The variable resistor in an ohmmeter circuit is used to compensate for
(A) Ac in the circuit (B) Improperly marked resistors (C) The incorrect polarity on semiconductor circuits (D) Circuit changes which will change the ohmmeter's current when checked with 0 Ω of unknown resistance

Answers to Self Tests

1. *D*
2. *C*
3. *A*
4. *C*
5. (a) 1.4 MΩ
   (b) 1.4 MΩ
   (c) 1.4 MΩ
   (d) 1.4 MΩ
   (e) 1.4 MΩ
   (f) 1 MΩ
   (g) 1 MΩ
   (h) 700 kΩ
   (i) 200 kΩ
   (j) 70 kΩ
6. 140 V
7. (a) 700 kΩ
   (b) 700 kΩ
   (c) 700 kΩ
   (d) 700 kΩ

   (k) 20 kΩ
   (l) 9 kΩ
   (m) 10 MΩ
   (n) 3 MΩ
   (o) 1 MΩ
   (p) 300 kΩ
   (q) 100 kΩ
   (r) 30 kΩ
   (s) 10 kΩ

   (e) 700 kΩ
   (f) 500 kΩ
   (g) 500 kΩ
   (h) 350 kΩ

   (i) 100 kΩ
   (j) 35 kΩ
   (k) 10 kΩ
   (l) 4 kΩ
   (m) 5 MΩ
   (n) 1.5 MΩ
8. (a) 0.10001 Ω
   (b) 1.001 Ω
   (c) 10.1 Ω
   (d) 111 Ω
9. 999.9 mA, 100 μA
10. So the meter will always be shunted when you are changing ranges.
11. (a) 4.985 MΩ
    (b) 1.485 MΩ
    (c) 485 kΩ
    (d) 135 kΩ
    (e) 35 kΩ

   (o) 500 kΩ
   (p) 150 kΩ
   (q) 50 kΩ
   (r) 15 kΩ
   (s) 5 kΩ

    (f) 0 Ω
    (g) 14,500 Ω
    (h) 1000 Ω
    (i) 15,000 Ω

12. 997 V
13. The 14.5-kΩ and 15-kΩ resistors decrease by the diode series resistance. Typically this is a few hundred ohms. The exact value depends on the diode used.
14. (a) 199,995 Ω
    (b) 19,995 Ω
    (c) 1995 Ω
    (d) 195 Ω
    (e) 15 Ω
    (f) 15 kΩ
15. 1995 Ω
16. *D*
17. *C*

VOM Specifications

■ This chapter discusses the specifications of the VOM. It is important for you to know the VOM's specifications in order to know the measurements it can make.

In this chapter you will learn the common VOM functions and the typical ranges associated with these functions. You will also learn to relate VOM accuracies to the cost of the instrument. In addition, you will learn how to use the VOM and some basic safety procedures to follow when using it.

4-1 INTRODUCTION

In Chap. 3 we saw how the meter movement is combined with resistance, switches, and a battery to build a VOM. We could only assume what the accuracy of any measurements would be. In most cases the manufacturer will not tell you the accuracy of the meter or the resistances being used. Therefore, you need a set of VOM specifications.

The VOM specifications tell you how well the VOM will do its job. The VOM specifications combine all the errors of the different parts used in one function. The error is then given to you as one number.

Like the specifications for any instrument, the VOM's specifications are very important. If you know the VOM's specifications, you will know the measurements it can make. You will also know what measurements are not possible because of the limits of the particular instrument. You will also know what measurements just should not be made by a VOM at all.

4-2 THE FUNCTIONS AND THEIR RANGES

Functions include such items as ac voltage, dc voltage, current, and resistance. As we noted earlier, most VOMs have the following four common functions:

1. Dc voltage
2. Ac voltage
3. Dc current
4. Resistance

A few VOMs have some extra functions. Some of these are temperature measurements (with a special probe), transistor parameter measurement, and capacitance measurement. Some manufacturers include decibel scales as an extra function.

One of the first questions you must answer when reviewing a VOM is "What are its ranges?" The VOM range specification is a basic description of the VOM's capabilities. For example, consider the VOM with three dc voltage ranges: 10, 100, and 1000 V. It is a much different instrument from the VOM with dc ranges of 1, 3, 10, 30, 100, 300, 1000, and 3000.

Fewer ranges make a less expensive VOM. A very low-cost VOM is often useful. However, it has real limitations. Because the number of ranges is quite limited, you must make some readings in the lower, less accurate portion of the meter scale. There may also be fewer low-end ac ranges than dc ranges, as the ac meter circuitry requires a minimum voltage before it works.

Meter ranges are chosen so you can make a reading in the upper portion of the meter scale. A number of VOM range sequences are common. For example, the range sequence 10, 50, 250, 500, 1000 allows all readings to be taken in the top 80% of the meter scale. You will also find the sequences 3–15–30 as well as the sequence 1–3–10. It is important to know the ranges you will need most often before you make the choice of a VOM.

Self Test

1. A VOM has a 1–3–10 sequence. What are the dc voltage ranges for measuring voltage from 1 to 3000 V?

2. If you need to read the 120-V ac power line accurately, which is the best range sequence for your job, and why?
   A. 10–25–50–500–1000
   B. 15–30–150–300–1500
   C. 10–30–100–300–1000
   D. 1–10–100

3. Which of the following is not a common VOM function?
   A. Dc volts
   B. Dc amperes
   C. Temperature
   D. Ohms

4-3 THE VOM'S ACCURACY

Generally, manufacturers classify VOM accuracy by the ac and dc functions. That is, you will find the VOM has one accuracy specification when making ac voltage measurements and another, better accuracy specification when making dc measurements. In most cases, you will not find an accuracy specification for VOM resistance-measurement circuits. Occasionally, you will find VOMs with even greater accuracy limitations. For example, very high current ranges or very low voltage ranges may have individual specifications.

Accuracy specifications are given as plus or minus (±) a percentage of full scale. The full-scale value depends on the range you have selected. For example, the accuracy of a VOM may be given as ±4% of full scale. This means that readings taken on the 30-V range are ±1.2 V (±4% of 30 V = ±1.2 V); readings on the 100-V range are ±4 V.

The accuracy of the VOM is determined by the accuracy of the meter movement used and the accuracy of the multipliers or the shunts.

Accuracies of ±1.5% for dc ranges and ±2% on ac ranges are exceptionally good. Accuracies of ±2% dc and ±3% ac are more common. Once in a while you will find VOMs with accuracies in the 4 to 5% region. Five percent is typical of a very low-cost product.

If ohmmeter accuracy is specified, it is normally specified in degrees of arc, as the scale is nonlinear. For example, an ohmmeter specification might read ±3° of arc. However, ohmmeters on VOMs are generally used to give an indication of what the resistance is, and not to give highly accurate resistance readings.

In order to fully understand the limitations of your meter, you must know both the accuracy and the range specifications.

For example, assume you wish to measure the 120-V ac power line. You have two VOMs available, a 5% meter with 15-, 30-, 150-, and 300-V ranges and a 3% VOM with 10-, 50-, 250-, and 500-V ac ranges. Which will do the best job of measuring the 120-ac power line?

The first VOM will be used on the 150-V range; its accuracy is

$$5\% \times 150 \text{ V} = \pm 7.5\text{V}$$

The second VOM will be used on the 250-V range, and its accuracy is

$$3\% \times 250 = \pm 7.5 \text{ V}$$

Both of these VOMs have the same actual accuracy when used on this range. Therefore, both will do the same job.

The accuracy of the ac voltmeter is usually specified only at one particular frequency. This is usually at 60 hertz (Hz), the normal power-line frequency. When the meter is used to measure frequencies other than 60 Hz, additional errors above the normal percentage of full-scale error are found. These may be specified in one of two ways. One common way is to specify the percentage of meter accuracy over a frequency range. This is usually the operating frequency range for the instrument. For example, the ac meter may be specified to have an accuracy of ±3% of full scale to 60 Hz, and additionally there will be no more than ±3% added error over the range of 20 to 1000 Hz. Another way the manufacturer may specify this accuracy is to give you a frequency range over which a VOM is accurate to within ±1 decibel (dB) (±1 dB = ±10%). Therefore, a meter with an accuracy of ±1 dB from 10 Hz to 10 kHz can be as much as 10% in error.

Self Test

4. A 9-V measurement is made on both the 10- and 50-V ranges of a ±3% VOM. What are the absolute accuracies of each measurement expressed as a percentage of the reading?

5. VOM accuracy specifications take into account inaccuracies in the
   A. Multiplier resistors
   B. Voltage source
   C. Diode characteristics
   D. Range switch contact resistance

6. An inexpensive VOM has an accuracy of ±5% of full scale. It is being used to measure the 120-V ac power line on the 1000-V ac range. What is the absolute accuracy expressed as ± a voltage? Expressed as ± a percentage of the reading?

7. A fully specified VOM ac voltage range includes the error at 60 Hz plus or minus
   A. The error at 1000 Hz
   B. The maximum error over its usable frequency range
   C. The error due to amplitude changes
   D. The error due to meter nonlinearities

4-4 THE VOM'S VOLTMETER INPUT IMPEDANCE

The input-impedance specification tells how much current the VOM will draw from the circuit you are measuring. Often this is called "loading" the circuit.

The VOM voltmeter input-impedance specification is given in ohms per volt (Ω/V). This specification is used because the impedance (resistance) changes as the voltmeter range changes. Looking at the schematic of Fig. 4-1, you can easily see that the meter resistance depends upon the range selected. If you selected a high voltage range, the load impedance of the voltmeter on the circuit is high. However, if you have selected the lowest possible range, the load impedance is the meter resistance plus a small multiplier resistor.

The most common dc input impedance for

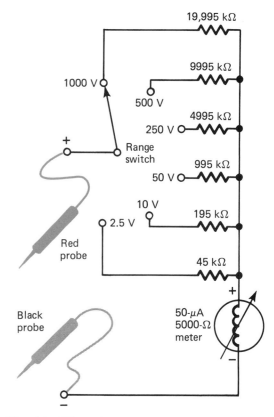

Fig. 4-1 The voltmeter circuit of a basic VOM.

VOMs is 20,000 Ω/V. The 20,000 Ω/V specification is derived from meters having a 50-μA sensitivity. By Ohm's law we can see

$$R = \frac{V}{I} = \frac{1 \text{ V}}{0.00005 \text{ A}} = 20,000 \text{ }\Omega$$

A 50-μA meter movement is quite common on better VOMs. If the VOM has a 1.5-V range, for example, the total input resistance is 30,000 Ω on this range.

$$R = 1.5 \text{ V} \times 20,000 \text{ }\Omega/\text{V} = 30,000 \text{ }\Omega$$

One of the more common ac input-impedance specifications is 1000 Ω/V. The 1000-Ω/V specification is chosen because of the ac meter/meter rectifier movement's characteristics. The 1000-Ω/V specification indicates the meter will draw 1 mA from the circuit under test, for a full-scale reading. One milliampere may be an insignificant current when drawn from the 120-V ac power line. However, it may become a significant current when drawn from a high-impedance circuit.

Alternatively, better VOMs are specified at

5000 Ω/V. These draw only 200 μA from the ac circuit which you are testing.

Very good VOMs specify the input capacitance on the ac ranges. The input-capacitance specification allows you to calculate the change in the load as the frequency of the ac signal is varied.

Self Test

8. A VOM has a 5000-Ω/V specification on its ac range. What current does it draw from the source when it is reading 30 V full scale? When it is reading 300 V full scale? When it is reading 150 V on the 300-V scale?

9. Figure 4-2 shows a 25-V voltage source with 100 kΩ of series resistance. You have a 20,000-Ω/V VOM with ±3% accuracy. Will the 30-V range or the 100-V range give you the best absolute accuracy expressed as a percentage of the reading?

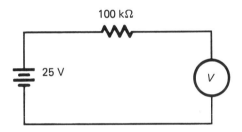

Fig. 4-2 Self test question 9.

10. VOMS with dc input impedance of 100,000 Ω/V are very difficult if not impossible to find. This is because
A. The 100,000-Ω/V VOM requires a 100-μA meter and 50-μA meters are more common for VOMs
B. The 100,000-Ω/V VOM requires a 10-μA meter, which is very difficult to obtain
C. The 1000-V range would have a 100-MΩ input impedance, which is difficult to build and very unnecessary

4-5 THE VOM'S CURRENT SHUNTS

Two VOM current-shunt specifications are important. The first specification is the overall ammeter accuracy. This is normally included with the dc accuracy specification. For example, a meter having a 1-A current range and ±2% of full-scale accuracy will read correctly on that range to within ±20 mA.

As indicated earlier, some shunts have special specifications. For example, a good VOM may have an overall ±2% dc specification. This specification may be difficult to keep for the 1- and 5-A shunts. As we have seen earlier, these shunts have a very low resistance value. High accuracies in the order of 1 or 2% are difficult to hold with very low resistance values. Therefore, it is not uncommon to see these high current ranges specified as having a ±3% of full-scale accuracy when the other current ranges have a ±2% of full-scale accuracy.

The other specification of interest answers the question, "How much voltage drop is required across the shunt when measuring currents?" This may be specified in one of two ways. Some manufacturers specify the value of the shunt resistance. Other manufacturers specify the maximum voltage drop across the shunt at full-scale deflection. This is often called the *insertion loss* of the meter. Of course, both of these specifications tell you the same thing. With either specification, some calculation is required to know the exact figures for your particular application.

For example, if a VOM has a 50-μA range and this range has a 20,000-Ω shunt resistance, what is the voltage drop when a current of 25 μA is being read? The answer is simple. Using Ohm's law, we see that

$$V = I \times R = 0.000025 \text{ A} \times 20,000 \text{ } \Omega = 0.5 \text{ V}$$

The manufacturer might have specified this range as 1 V drop at full scale. As 25 μA is one-half scale, we know the voltage drop is 0.5 V.

Both the input-impedance and the shunt-drop specification tell you how much you will disturb the circuit when you make the measurement.

The effect of voltage drop across the shunt resistor will depend on the circuit in which you are taking the measurement. There will be little or no effect in some circuits, and in others the measurement of current will change the current being measured. Usually, you can determine this quite easily by examining the voltages of the circuit in which you are going to make a current measurement.

For example, suppose you are going to make a current measurement in a circuit whose total voltage drop is 3 V. You are

37

using a VOM whose ammeter causes a 0.3-V insertion loss. In this particular situation the 0.3-V insertion loss caused by the ammeter is 10% of the total circuit voltage. It is easy to see that this will change the total current flowing in the circuit.

On the other hand, suppose you were going to measure a current in a circuit using a 60-V power supply. In this situation, the 0.3-V insertion loss amounts to only ½% of the total circuit voltage. Again you can easily see this will cause little change in the total circuit current.

Self Test

11. A 15-V source develops a 100-mA current through a 150-Ω load. An ammeter with 0.25-V insertion loss is used to measure the current. What error is caused by this meter? How would the error change if the current was generated by a 150-V source and a 1500-Ω load?

12. You are using a VOM which has a 100-μA full-scale current range. The manufacturer's specifications tell you that the shunt resistance is 10,000 Ω on this range. What insertion-loss voltage could you expect from this ammeter when measuring a full-scale current?

13. Your VOM has a 3-A dc current range. The manufacturer has specified that this ammeter has a maximum 0.3-V insertion loss. By Ohm's law you have calculated that this ohmmeter has approximately a 0.1-Ω shunt on this range. When you make the connection to the circuit, you introduce an extra 0.1-Ω resistance where the test leads make connection to the circuit. How will this affect your current measurement?

4-6 SOME SPECIAL SPECIFICATIONS

VOMs normally specify the center-resistance reading. For example, a common specification is 10 Ω center resistance in the $R \times 1$ position. If the meter is used in the $R \times 10$ position, the meter reads 100 Ω center scale. In the $R \times 100$ positions it reads 1000 Ω center scale.

Frequently, additional special scales are supplied with the VOM. For example, scales may be calibrated in decibels, VU (volume

units), or dBm. The abbreviation dBm stands for decibels referenced to 1 mW.

In addition to the electrical specifications, you should also look at the VOM's mechanical specifications. These tell you the size of the meter and the temperature range for operation. Power requirements indicate the batteries required to power the ohmmeter circuits.

Occasionally you will find specifications indicating the durability of the instrument. Some VOMs have been designed for fairly rugged use. These units can be dropped or otherwise misused and will still remain operational. *Note:* Operational does not mean that these VOMs are still in calibration. Once a VOM has been dropped or has received any severe blow, you should check its calibration. Periodic maintenance of your VOM will certainly go a long way toward improving its life.

4-7 USING THE VOM

Most VOMs are quite simple to use. The most negative input signal is connected to the common jack. This jack is connected to the circuit with a black test lead. The most positive input signal is connected to the jack marked V-Ω-mA, +, or volts-ohms-mA. This jack is used for most voltage and current measurements. It is used for all resistance measurements.

High-current and high-voltage measurements are often made between the common input and special jacks just for this purpose. These special jacks are used because the range switch cannot carry the voltages or currents required by these special ranges.

Measurements should be made starting at the highest range first. The meter can then be down-ranged until the pointer is as far upscale as it can go without being off scale. For example, suppose you have a VOM with a 10–50–250–500–1000-V range sequence. To measure a 100-V source, the meter is first set on the 1000-V position. The reading is then at one-tenth of full scale. The range switch is then set to the 500-V position. The meter then reads one-fifth full scale. The VOM range switch is then set to the 250-V position. At this time the meter reads two-fifths of full scale. If you move the range switch to the next position, the meter will go off scale. The reading is therefore made on the 250-V range.

The VOM is a relatively safe instrument.

However, some safety precautions should be observed.

When the VOM is being used to measure high voltages, the other front-panel jacks may present a shock hazard. You should be sure to keep your fingers out of contact with these jacks.

The test-lead tips also offer a possible shock hazard. If you accidentally touch the probe tips while making a measurement, you are inviting a shock. Of course, you also should be very careful of the test leads themselves when measuring extremely high voltages. If there is any defect in the insulation, you may get a shock from the test lead itself.

If one of the test probes falls off the circuit being tested, the other lead may easily be connected to a high-voltage source. This leaves the entire VOM, *including* the loose test probe, at this high voltage. *Note:* This can happen during current measurements as well as voltage measurements.

Probably the greatest source of shock while using the VOM comes when the user does not pay attention to how the measurement is being made. Simple carelessness can lead to many unnecessary shocks. Remember, more fatal shocks occur from the 120-V ac power line than any other source. Many times they are simply due to carelessness!

14. Describe the procedure you would use when measuring the dc high voltage which might be found in a tube-type television set.

15. You are measuring a 250-V dc supply. The common test lead slips off the chassis. What could happen if you accidentally grab the common test lead by the probe tip as you are attempting to reconnect it to the chassis?

16. Your VOM ohmmeter is set to the $R \times 100 \ \Omega$ scale. This particular meter has a 20-Ω center scale. Instead of making a resistance measurement, you accidentally connect the meter across 110 V ac. What might happen to the meter? What kinds of currents are drawn? What voltages are produced across different parts of the meter?

17. You are using your VOM to measure circuit resistances. You have set the range switch to the $R \times 1000$ position. You are attempting to measure what should be a 22-kΩ resistance, but the red test lead is not connected to the V-Ω-mA jack. Instead, it is connected to the +4000-V jack. What damage will be caused to the instrument? Why?

Summary

1. The function and range specifications tell you what the VOM does and over what range it will do it.

2. The most common functions are ac and dc volts, dc current, and resistance.

3. The range gives the highest possible value which can be read at that setting.

4. VOM accuracy usually depends on the function. The accuracies for ac measurements are usually poorer than those for dc measurements. Accuracies are given as a percentage of full scale.

5. The VOM input-impedance specification tells how much the voltmeter loads the circuit under test. The resistance of the VOM input changes with each new range setting. The VOM input impedance is therefore given in ohms per volt.

6. There are two important specifications about VOM ammeters' overall accuracy and voltage drop. The voltage drop in the circuit being tested can be specified in two ways. The manufacturer can give the insertion (shunt) resistance on each range or specify the voltage drop on all ranges.

7. The VOM is a very simple instrument to use. It should be connected to the circuit when the circuit is off.

8. You should start with the highest possible range-switch setting and then switch to lower and lower values until a usable meter reading is found. In order to avoid electrical shock, you should make a practice of keeping your hands away from probe tips and the test-lead jacks.

4-1. Two VOMs have 1- to 1000-V ranges. One has a 1–10–100 range sequence and the other has a 1–3–10 range sequence. The VOM with the 1–3–10 range sequence lets you
(A) Make more accurate readings at exactly 1000 V (B) Make more readings in the upper two-thirds of the meter scale (C) Use the instrument in 120/240-V ac applications (D) Make power measurements

4-2. If you have a VOM with a 1–2.5–5 range sequence and it is set on the 50-V range, the next most sensitive range is
(A) 100 V (B) 50 V (C) 25 V (D) 10 V

4-3. VOM voltmeter accuracies are usually given as
(A) A percentage tolerance for the multiplier resistors (B) ± a percentage of reading (C) ± a percentage of full scale

4-4. You use a ±4% VOM on the 500-V range to measure a 440-V power source. What is the absolute accuracy expressed as ± a voltage? Expressed as ± a percentage?

4-5. The accuracy specification for the VOM's ac voltmeter must include
(A) The meter-rectifier error at 60 Hz (B) The multiplier-resistor error (C) The meter-rectifier error at the operating frequency (D) All of the above

4-6. The input impedance for a VOM is specified in ohms per volt because
(A) It is easier to use (B) The input impedance changes with each range setting (C) No two VOMs are exactly alike (D) Most VOMs use a 50-μA meter

4-7. Your new VOM uses a 50-μA meter movement. You would expect the dc input impedance to be
(A) 1000 Ω/V (B) 5000 Ω/V (C) 20,000 Ω/V (D) 50,000 Ω/V

4-8. Typically the input impedance of your VOM's ac voltmeter is much lower than that of your dc voltmeter. A value of ___?___ is typical for a good instrument.
(A) 1000 Ω/V (B) 5000 Ω/V (C) 20,000 Ω/V (D) 50,000 Ω/V

4-9. You are using your 20,000-Ω/V VOM to measure 150 V dc. It is set on the 300-V range. The voltmeter draws ___?___ from the circuit under test.
(A) 25 μA (B) 50 μA (C) 150 μA (D) 300 μA

4-10. The VOM ammeter specifications tell you two things. First, they tell you the full-scale current and second they tell you
(A) The ac frequency response (B) The current drawn from the circuit under test (C) Whether the error is ± a percentage of full scale or of the reading (D) The insertion loss

4-11. A high insertion loss has the greatest effect on circuits with a
(A) High source voltage and high current (B) High source voltage and low current (C) Low source voltage and high current (D) Low source voltage and low current

4-12. Often the insertion loss on the highest current range is greater than the insertion loss on all the other ranges. Why?

4-13. You are using a VOM with a 10-Ω center-scale ohmmeter. You are using it on the $R \times 10,000$ range. How much will the pointer move when you change the unknown resistor from 1 to 10 MΩ?

Answers to Self Tests

1. 1 V, 3 V, 10 V, 30 V, 100 V, 300 V, 1000 V, and 3000 V.
2. *B*, because the 150-V range allows the closest to full-scale 120-V measurement.
3. *C*
4. ± 3.33% and ± 16.7%
5. *A*
6. ± 50 V and ± 41.7%
7. *B*
8. 200 μA on the 30-V scale and 200 μA on the 300-V scale. 100 μA when reading 150 V on the 300-V scale.
9. The 30-V range has a 600-kΩ input impedance. Using the 30-V range, the total load on the source is 700 kΩ (600 kΩ + 100 kΩ). This causes 35.7 μA to flow, giving a drop of 3.57 V across the 100-kΩ internal source resistance. Therefore, the reading could be 3.57 + 0.9 V, or 4.47 V low.

   The 100-V range has a 2-MΩ input impedance. Using the 100-V range, the total load on the source is 2.1 MΩ

   (2 MΩ + 100 kΩ). This causes 11.9 μA to flow, giving a drop of 1.19 V across the internal source resistance. Therefore, the reading could be 1.19 V + 3.0 V, or 4.19 V low.

   Although the two will be close, the 100-V range will be out no more than 16.8%, whereas the 30-V range could be out 17.9%.
10. *B*
11. The source voltage is dropped to 14.75 V. Therefore, the current drops from 100 to 98.3 mA. This is 1.7% error. Using the higher voltage, the current drops only to 99.8 mA. This is only a 0.2% error.
12. 1 V
13. It will not change the reading unless the extra 0.1-Ω contact resistance actually changes the current source. It will not change how the meter reads because the probe resistance is not part of the shunt.
14. (*a*) Set the VOM to the

highest-range switched value. (*b*) Clip the common lead to the chassis. (*c*) If possible, with the TV turned off, clip the + probe to the point to be measured. (*d*) Measure the voltage. (*e*) Change to a lower range if possible.
15. You could get a 250-V shock through the VOM. The current will be limited to 50 μA or less by the series resistance of the VOM's multipliers.
16. Unless the meter movement is diode-protected, it will probably be burned out by the 120 V ac applied directly across the meter and its 1.5-V multiplier. This gives roughly a 100 × current overload in the meter. The 2000-Ω calibration resistor will have a 60-mA current. The 60 mA at 120 V will dissipate 7.2 W and destroy this part.
17. There will be no damage to the instrument because no current will flow. You will just get no reading.

41

The Electronic Meter

■ This chapter discusses the analog electronic meter. This meter can be used to make measurements that cannot be made with the VOM.

In this chapter, you will learn the specific reasons for using an electronic meter rather than a VOM. You will also become familiar with the differences between various types of electronic meters and learn to draw block and schematic diagrams of electronic-meter circuits.

5-1 INTRODUCTION

We looked at the VOM in Chaps. 3 and 4. We found it is a practical tool for measuring dc voltage, ac voltage, dc current, and resistance. Using the VOM, you can measure many voltages, currents, and resistances found in electric and electronic circuits. Unfortunately, the VOM cannot make all the measurements we wish to make. For example, most VOMs will not measure ac current. This is a frequently needed function. We know that most VOMs use a 50-μA meter movement. This means the VOM with a full-scale voltage reading draws 50 μA from the circuit under test. Many times this "loading" will disturb a sensitive electronic circuit. Sometimes a circuit will fail to work because of meter loading. At other times it will cause your measurements to be in error.

We also observed that the VOM is not very sensitive. Very good VOMs give only 0.25-V full-scale dc ranges and typically 1.5-V full-scale ac ranges. Frequently these are not sensitive enough for work in modern electronic circuits. This is especially true with solid-state equipment.

The electronic meter uses vacuum-tube or transistor amplifiers. These "active devices" are used to build a meter with better features and specifications than the VOM. Throughout this chapter we will call the an-

alog electronic meter simply the electronic meter. In Chap. 7 we look at the digital electronic meter.

5-2 ADVANTAGES OF THE ELECTRONIC METER

As noted previously, the electronic meter corrects some of the VOM's problems. One of the major problems with the VOM is the amount of current it draws from the circuit when measuring voltages.

Any current drawn from the circuit changes the way the circuit operates. This change becomes a reading error. The perfect voltmeter draws no current from the circuit under test, but no voltmeter is perfect. The electronic meter uses active devices to make a close to perfect instrument. Typically, VOMs require 50 μA for full-scale readings. This current requirement is expressed by the VOM's ohms per volt specification.

We calculated the 1.5-V range input resistance on a typical 20,000 Ω/V VOM. We found it is 30,000 Ω. One of the important features of the electronic meter is that it has the same high input impedance on each voltage range. Most electronic meters have a 10-MΩ input impedance.

We know that 20,000 Ω/V with a 1.5-V scale draws 50 μA from a 1.5-V source. How much current does the 10-MΩ electronic voltmeter draw from a 1.5-V source? By Ohm's law, we find

$$I = \frac{V}{R} = \frac{1.5\ V}{10 \times 10^6\ \Omega} = 0.15 \times 10^{-6}\ A$$
$$= 0.15\ \mu A$$

Obviously, this is much less circuit loading than the VOM causes.

Electronic meters have a constant input impedance on all voltage ranges. That is to say, the input impedance is the same on each setting. Therefore, at 1000 V, the circuit loading is

$$I = \frac{V}{R} = \frac{1000\ V}{10 \times 10^6\ \Omega} = 100 \times 10^{-6}\ A$$
$$= 100\ \mu A$$

However, the VOM loads the circuit at 20,000 Ω/V. Therefore, on the 100-V range with 1000 V input, the circuit loading is

$$20,000\ \Omega/V \times 1000\ V = 20 \times 10^6\ \Omega$$
$$\text{and} \quad I = \frac{V}{R} = \frac{1000\ V}{20 \times 10^6\ \Omega} = 0.00005\ A$$
$$= 50\ \mu A$$

As you can see, on the highest voltage ranges, the VOM may not load the circuit as much as the electronic voltmeter does.

The electronic meter is built using vacuum-tube or transistor amplifiers. The amplifiers make the instrument more sensitive than its meter movement alone will allow. The best VOM sensitivity is normally 0.25 V full scale. The electronic meter gives these instruments 0.1 to 0.001 V full-scale sensitivity. This added sensitivity is very useful, especially for semiconductor circuits.

Let's look at an example. We are using an electronic meter on the 10-mV scale. The meter has a 10-MΩ input impedance. When we are measuring a 10-mV signal, the input current can be calculated by Ohm's law. The current is

$$I = \frac{V}{R} = \frac{0.01\ V}{10 \times 10^6\ \Omega} = 1 \times 10^{-9}\ A = 1\ nA$$

A 1-nanoampere (1-nA) current is very small. We can see what the electronic amplifier is adding to this instrument. The meter movement, for example, is 1 mA full scale. A 1-nA

current is therefore amplified to 1 mA. The gain is

$$\text{Gain} = \frac{\text{output current}}{\text{input current}}$$
$$= \frac{1000\ \mu A}{0.001\ \mu A} = 1,000,000$$

We have gained two things. First, we have more sensitivity than we could ever have had. They don't make 1-nA full-scale meter movements! Second, the electronic meter can use a much less expensive 1-mA meter movement instead of the expensive 50-μA movement we needed to build a good VOM.

The increased sensitivity shows up even more on the ac voltage ranges. Typical maximum VOM ac sensitivities of 1.5 V full scale are common. Using electronic amplification, the sensitivity of the ac and dc ranges can be the same.

With the electronic amplifier we add special features not found on VOMs. For example, we can place a battery in series with the meter movement. Because there is an amplifier between the meter and the input terminals, this only changes the reading. Now we can make 0 V at the input cause the meter pointer to rest at center scale. Positive input voltages cause the pointer to deflect toward full scale. Negative input voltages cause the pointer to deflect toward zero, or the left-hand part of the scale. This useful feature can be added electronically to the meter using tubes or transistors.

Some electronic meters include automatic polarity detection and correction. Positive or negative signals do not require you to make a function change. The meter pointer always reads up scale. Polarity indicators show whether the signal is positive or negative.

Self Test

1. The electronic meter is different from the VOM because it is built using electronic parts that are
   A. Active
   B. Passive
   C. Resistive
   D. Capacitive
   E. Inductive

2. You have both a 20,000-Ω/V VOM and a 10-MΩ electronic meter. Both instruments have 1-, 2.5-, 5-, 10-, 25-, 50-, 100-,

From page 42:
Input resistance

On this page:
Circuit loading

Constant input
impedance

Sensitivity

Zero adjust

Automatic
polarity

Vacuum-tube
voltmeter

Transistorized
voltmeter

250-, 500-, and 1000-V ranges.   Complete the following table:

Range, volts	VOM		Electronic	
	Input *I*	Input *Z*	Input *I*	Input *Z*
1				
2.5				
5.0				
10				
25				
50				
100				
250				
500				
1000				

3. Refer to your completed table from question 2. At what range setting do both the VOM and the electronic meter have the same input impedance? On this range do these instruments appear any different relative to the circuit being tested? Why?

4. The electronic meter offers better sensitivity than the VOM. The best sensitivity improvement is found when comparing the _____?_____ scales.
   A. Ac voltage
   B. Dc voltage
   C. Dc current
   D. Resistance
   E. Ac current

5. The addition of _____?_____ circuits allows special features to be added to the electronic meter.
   A. Active
   B. Passive
   C. Resistive
   D. Vacuum-tube
   E. Transistor

5-3 FIVE ELECTRONIC METERS

When we speak of electronic meters, we are talking about a large group of voltage-, current-, and resistance-measuring instruments. When we call them electronic meters, we are saying they use vacuum-tube or transistor amplifers to improve the instrument's performance. Vacuum tubes or transistors are used to overcome many of the limitations we found in the simple VOM.

If vacuum tubes or transistors are used, the electronic meter must have a power supply. The power supply can be run from batteries or from the ac power line. This means the electronic meter is not as portable as the VOM. Operation from the power line means we must be able to plug the electronic meter in. Of course, this restricts its portablity quite a bit. Batteries, as we know, have a limited life. We also know that adding transistors or vacuum tubes to an instrument must increase its cost. The extra parts also mean the instrument may require more frequent service. From this discussion we can see that we do not get something for nothing. When we gain the advantages of the electronic amplifier, we give up some of the portability, simplicity, and low price.

A number of different electronic meters are in use today. Most of them do almost the same thing, but each one has its own name describing a slightly different job. We will review these different types.

The VTVM

The VTVM was the first of the electronic voltmeters. The initials VTVM stand for vacuum-tube voltmeter. Although no new VTVMs are being designed today, a great number of VTVMs remain in service shops, in laboratories, and on experimenter's benches. The VTVM is a very reliable instrument. The vacuum tube is an extremely rugged device and is much harder to damage with electrical overloads than the transistor is. The circuits in the VTVM are very simple, and therefore many old VTVMs are still in daily use. VTVMs usually measure only ac voltage, dc voltages, and resistance. Because nearly all VTVMs are ac-power-line-operated, it is not usually considered a portable instrument.

The TVM

The TVM (transistorized voltmeter) is the modern replacement for the VTVM. In many cases, TVMs are simply transistorized versions of earlier VTVM designs. Some of the newer TVMs have features not found in the VTVMs. For example, some newer TVMs have current ranges, greater sensitivity, and LED polarity indicators. Some TVMs operate from the power line, others operate from internal batteries, and some can use either power source. The TVM is easier to damage with an electronic overload than the VTVM is. This is because transistors are not as rugged as vacuum tubes. Battery-operated TVMs are much more portable than either line-operated VTVMs or line-only-operated TVMs.

The FETVOM

The FETVOM (field-effect transistor volt-ohmmeter) is a specialized version of the TVM. As its name tells you, the FETVOM is built with field-effect transistors (FETs). The FET is used as the input transistor to the meter amplifier. This is because it makes an amplifier with a very high input impedance. This high input impedance gives us meters which do not load circuits as much as VOMs. FETVOMs include all the voltage and current ranges normally found on VOMs. Often FETVOMs also include ac current-measurement capabilities. They also have higher input impedance and greater sensitivity than the VOM. Most FETVOMs are operated from an internal battery supply. The names FETVOM and TVM are often incorrectly used for one another. These instruments are different, however, each having its own distinct characteristics.

The ACVTVM and the ACTVM

The ACVTVM and the ACTVM are special versions of the vacuum-tube voltmeter and the transistorized voltmeter. These special electronic meters are designed to measure ac voltage only. You can easily recognize the ac voltmeter by its lack of a dc voltage function, current function, or resistance function. You will also notice much more sensitive voltage ranges than the voltage ranges you will normally find on other electronic meters or VOMs.

For example, the most sensitive range on a good TVM might be 0.1 V full scale. This would be the same for both the ac and the dc voltage functions. Looking at a typical ACTVM, we find the most sensitive range is 1 mV full scale. We may also find other features such as scales calibrated in decibels and special outputs to drive chart recorders. You will use ac electronic meters most often when you are working with audio equipment. The ac electronic meters have another feature. They usually are able to measure voltages over a wider frequency range, while still maintaining their accuracy, than are other types of electronic meters. This feature is called *extended frequency response*.

In most cases, the ACTVM is simply a transistorized version of the ACVTVM. Occasionally you will find the ACTVM has a few special features not found on earlier ACVTVMs.

Self Test

6. An electronic meter is
   A. A meter used only on electronic circuits
   B. A meter useful only on electronic circuits
   C. A meter containing electronic circuits
   D. A meter which cannot be used to measure the ac power line

7. TVM means ____?____ voltmeter.
   A. Transistorized
   B. Tube
   C. Transportable

8. Most TVMs operate from
   A. No power source
   B. Line voltage
   C. Both line voltage and batteries

9. The VTVM is the ____?____ electronic voltmeter designed.
   A. Most recent
   B. Earliest
   C. Most accurate
   D. Most versatile

10. Ac electronic voltmeters are used primarily by
    A. Electricians
    B. Audio technicians
    C. Power-generating stations
    D. Television stations

11. The VTVM would normally be considered a ____?____ test instrument.
    A. Bench
    B. Portable
    C. Disposable
    D. Rack-mount

12. The FETVOM is so named because
    A. It is used to test FETs
    B. It will not harm FETs when testing them
    C. Field-effect transistors are used in its circuits
    D. We need to identify the active versus the passive VOM

13. The electronic amplifier can be used to ____?____ the sensitivity of a meter movement.
    A. Increase
    B. Decrease
    C. Offset
    D. Reverse

45

14. The input impedance of an electronic voltmeter ____?____ as the voltage range is increased.
    A. Increases
    B. Remains the same
    C. Decreases

15. At 400 V, the ____?____ has the greatest input impedance.
    A. 20,000-Ω/V VOM
    B. 10-MΩ TVM
    C. 1-MΩ TVM
    D. 5000-Ω/V VOM

16. The input impedance of the 20,000-Ω/V VOM in self test question 15 is ____?____ .

5-4 ELECTRONIC VOLTMETER CIRCUITS

Figure 5-1 shows a diagram of the basic components of an electronic analog multimeter. These are:

- Meter amplifier
- Input attenuator
- Rectifier
- Meter movement
- Ohmmeter circuits

A form of each of these elements or blocks will be found in the five basic electronic meters discussed in Sec. 5-3. If the instrument does not have the ohmmeter function, the ohmmeter circuits will not be used. The order of connecting these blocks may change. For example, most VTVMs rectify the incoming signal before the meter amplifier, but most TVMs rectify the incoming signal after the meter amplifier.

As you look at the basic diagram in Fig. 5-1, one difference between the electronic meter and the VOM becomes clear. The common terminal of the electronic meter is connected to its own power supply. Often this power supply is connected to the 120-V ac power line. In many cases this means the common test lead and the power-line ground are electrically connected. For this reason most electronic meters have a dc polarity switch. The polarity switch lets you read either positive or negative dc signals as up-scale deflections. This is done because you cannot simply reverse the test leads to make a negative-polarity measurement, as you did with the VOM. If you reverse the test leads, you will connect the grounded common lead to the negative voltage source. This will short the source to ground.

On some instruments this is done automatically. This feature is called *autopolarity*. The instrument has an indicator to tell you which polarity voltage or current you are measuring.

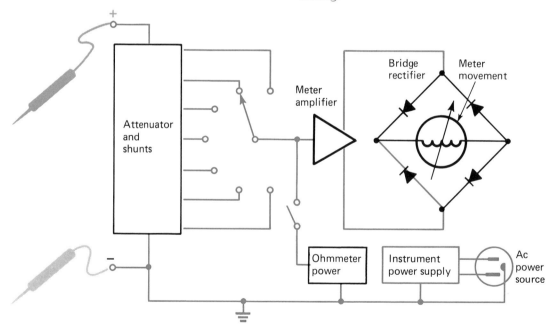

Fig. 5-1 A diagram of a typical electronic analog multimeter. All new designs place the rectifier in the amplifier or around the meter. Early **VTVM** design sometimes rectified ac signals before they are amplified.

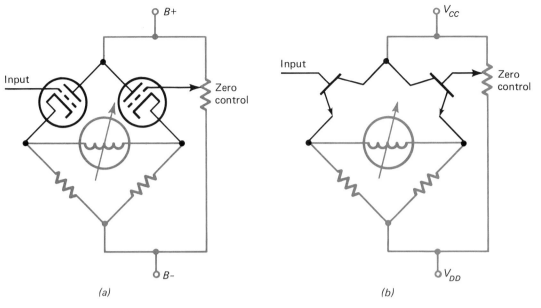

Fig. 5-2 Meter amplifiers. (*a*) Vacuum tube and (*b*) transistor. Both amplifiers are connected as followers in the arm of a bridge. The input impedance of the amplifier is as high as the input impedance of the cathode or emitter follower.

Impedance buffer

Gain

Operational amplifier

Zero control

We will now look at the circuits for each of these blocks individually.

The Meter Amplifier

The basic building block which makes the electronic meter different from the VOM is the meter amplifier. The meter amplifier serves two purposes.

First, it makes the input impedance or "loading" of the meter independent of the meter movement used. This means the loading an electronic voltmeter puts on the circuit depends only on the amplifier and the input attenuator. It does not depend on the resistance of the meter movement itself.

Second, the amplifier is used to greatly increase the meter sensitivity. For example, we know the typical VOM requires 50 μA to deflect the pointer to full scale. Using an amplifier with a current gain of 50, only 1 μA is needed to deflect the pointer to full scale.

Figure 5-2 shows a typical meter-amplifier circuit in both vacuum-tube and solid-state forms. This circuit is a bridge using active devices as two of the four arms of the bridge. These active devices are used in the follower mode. The amplifiers provide no voltage gain. However, they do have a great deal of current gain. This means that they do increase the input impedance.

The circuits of Fig. 5-3 are used to provide both impedance buffering and gain. These circuits do not have as high an input impedance or the stability of the circuits shown in Fig. 5-2. Again, we see we do not get something for nothing. However, these circuits do give voltage gain. Therefore, the meter using these amplifiers has much higher voltage sensitivity as a major advantage.

Many modern electronic-meter designs do not use the transistor or the FET in a bridge circuit. Instead, the integrated-circuit operational amplifier is used. The operational amplifier is often used with FET followers, as shown in Fig. 5-4. The use of the FET followers gives the very high input impedance that is needed. The integrated-circuit operational amplifier gives the controlled gain desired.

The electronic voltmeter offers one feature which is not found on a simple VOM. This feature is the zero control. The zero control allows you to add or subtract a dc voltage from the displayed meter reading. You can add or subtract the full-scale voltage from any input signal. The zero control does the same job as the mechanical-meter zero adjustment. The electronic zero control just has a much wider range than the mechanical adjustment.

Most of the time the zero control is used to place the zero voltage (or current) position at center scale. When you do this, positive

47

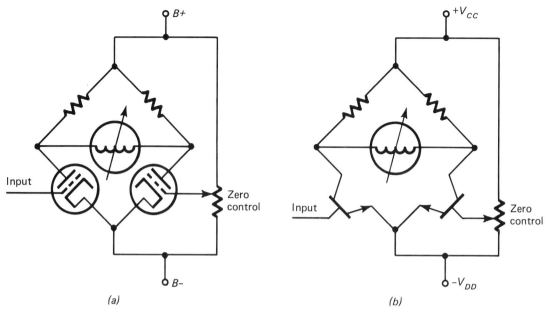

Fig. 5-3 Two simple meter amplifiers with gain. (*a*) Vacuum
tube and (*b*) transistor. Although the bridge configuration is
still used, these circuits give the input-signal voltage gain as well
as increased input impedance.

signals move the meter pointer up scale from
center, and negative signals move the meter
pointer down scale from center.

The current flows in Fig. 5-5 show how the
zero control on a bridge amplifier works. As
you can see in Fig. 5-5(*a*), both of the FET

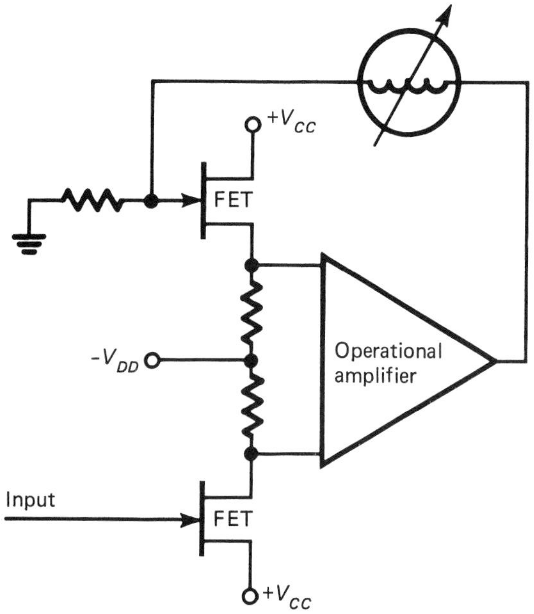

gates are at the same voltage. In this case
they are both zero volts. When the inputs are
the same voltage, identical currents pass
through both sides of the bridge. No current
passes through the meter; therefore, it reads
zero.

In Fig. 5-5(*b*) the gate of Q_2 has gone negative. The "zero" side still has no input. That
is, it is at zero volts. Transistor Q_1 is turned
on more than transistor Q_2. Part of the current that was passing through Q_2 is now di-
verted and passes through the meter and then
through Q_1. Because the current is flowing
through the meter, the meter is now deflected
up scale.

Self Test

17. The purpose of a meter amplifier is to
    A. Reduce meter drift
    B. Give the meter impedance buffering
    C. Improve ac response
    D. Provide an expanded resistance func-
    tion

18. A meter amplifier with voltage gain allows
    you to build an instrument with
    A. A lower-cost meter movement
    B. Greater sensitivity than the meter
    movement alone
    C. Higher input impedance
    D. Drift-correction circuitry

Fig. 5-4 The FET input operational amplifier
used as a meter amplifier. The operational
amplifier is used in the follower with gain mode
to give an extremely high input impedance and
the required gain.

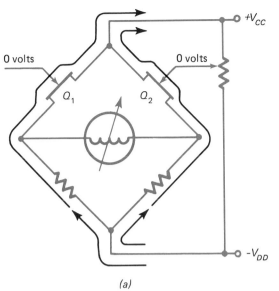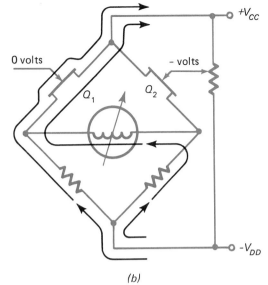

0 volts   0 volts   +V_{CC}

Q_1 Q_2

−V_{DD}

(a)

0 volts   − volts   +V_{CC}

Q_1 Q_2

−V_{DD}

(b)

Dividing the
input signal

Resistive voltage
divider

Fig. 5-5   Current flow in the bridge amplifier. (*a*) The current
is balanced between both sides of the amplifier. (*b*) Adjusting
the zero control to cut off Q_2 forces current through the meter
and Q_1.

19. The follower type of bridge amplifier gives
_____?_____ the meter movement.
A. The same sensitivity as
B. Much more sensitivity than
C. One-half the sensitivity of
D. Twice the sensitivity of

20. The zero control lets you add to or sub-
tract from the displayed reading a voltage
equal to the
A. Instrument's $B+$ supply
B. Instrument's $B-$ supply
C. Instrument's input voltage
D. Full-scale voltage

21. The meter and the meter amplifier set the
A. Basic instrument sensitivity
B. Maximum working voltage
C. Maximum working current
D. Maximum working resistance

Input Attenuators

In any electronic meter, the amplifier/meter-
movement combination sets the maximum
sensitivity of the entire instrument. They
make the instrument as sensitive as the instrument will ever be. If you need less sensitivity,
the input signal must be divided. This is done
by sending the signal to the input attenuator
before it goes to the amplifier in the meter.

For example, suppose the FET operational-
amplifier circuit of Fig. 5-4 is used, and the
voltage gain of the amplifier is set at 100. The
meter movement used is a 1-mA 10,000-Ω
movement. By Ohm's law, we can see that
the voltage across the meter is 10 V when a
full-scale current flows:

$$V = IR = 0.001 \text{ A} \times 10,000 \text{ } \Omega = 10 \text{ V}$$

The amplifier at a gain of 100 means only
10 mV is required at the input to give a full-
scale reading. The combined amplifier/meter
sensitivity is always 100 mV full scale. If you
wish to measure a 1-V signal, it must first be
divided by 10.

For all practical purposes, meter amplifiers
have infinite input impedance. That is, they
draw no current from their input. The infinite input impedance lets us use a different
method to change voltage ranges for the electronic meter than we used in the VOM. Figure 5-6 shows the difference between a VOM
voltage-range switch and an electronic-meter
voltage-range switch.

The VOM uses a different value of multiplier resistor for each voltage range. The
electronic meter puts the input voltage on a
resistive voltage divider. The range switch selects the needed amount of division. The resistor values used in the divider are picked for
a meter amplifier with infinite input impedance. Anything less than an infinite imped-

49

Input impedance

Ac attenuator

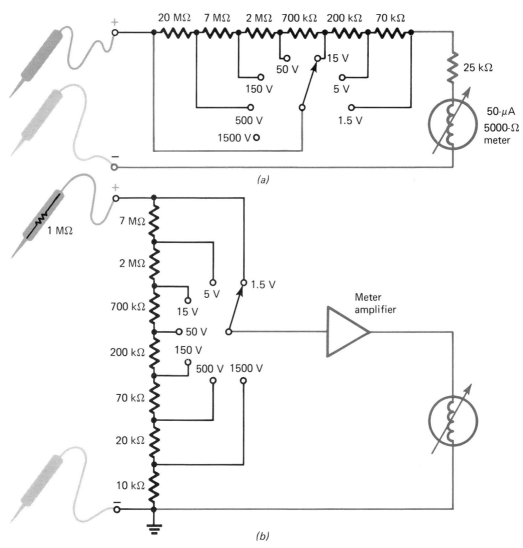

Fig. 5-6 (a) The VOM range selector. (b) The electronic-meter input attenuator. The VOM range adjustments are simply increasing or decreasing meter multiplier resistors as needed. The electronic-meter attenuator uses a voltage divider to reduce the input signal to the range selector.

ance will load the divider. This will make the division ratio wrong.

The divided output is connected to the meter-amplifier/meter-movement combination. Remember this has fixed sensitivity. If a voltage divider is used, the load placed on the circuit being measured is always the same. It does not change with the selected voltage range.

Most electronic meters are designed with a total voltage-divider resistance of 10 MΩ for dc voltage measurements. Typical resistance values as shown in Fig. 5-6 for a common dc 10-MΩ attenuator. VTVMs have an additional 1 MΩ at the probe tip (see Fig. 5-6). This raises the total VTVM input impedance

to 11 MΩ. It is possible to have a common input attenuator for both the ac and dc measurements. However, separate attenuators are sometimes used because the high-frequency characteristics of a 10-MΩ voltage divider are difficult to control. For this reason, the total resistance of some ac attenuators is only 1 MΩ.

Figure 5-7 shows a typical ac input attenuator. Notice the resistive divider is paralleled by a capacitor divider. The perfect resistive divider would be completely free of any stray capacitance. This means there would be no requirement for the additional "compensating" capacitors. Unfortunately, real resistive dividers do have stray capacitance.

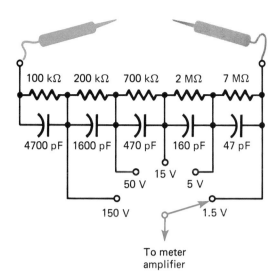

Fig. 5-7 The compensated attenuator. The values of capacitance placed across the resistive divider are high enough to swamp out the effects of any stray capacitance present in the divider resistor.

The stray capacitance is not the same ratio as the resistance ratios. For this reason, ac signals will be divided by a ratio which depends on the resistive ratios and the stray-capacitance ratios. To prevent this from happening, additional capacitance is added. This makes the capacitance and the resistance ratios equal. This is called "compensating" an attenuator. The compensating capacitors are large enough to swamp out the effects of any stray capacitance.

The addition of these capacitors means signals at the input look at both a resistance and a capacitance. The capacitance is the series value of all the compensating capacitors plus the value of any stray capacitance.

Self Test

22. The basic purpose of the input attenuator is to _____?_____ the input signal.
    A. Divide
    B. Amplify
    C. Correct
    D. Compensate

23. The input attenuator is needed because
    A. The meter amplifier cannot handle extreme overloads
    B. The meter multiplier resistors will not do the complete job
    C. The input amplifier and the meter combination has a limited signal range

D. Input attenuators are simpler to build than multipliers

24. The typical dc attenuator has a total resistance of
    A. 20 MΩ
    B. 10 MΩ
    C. 1 MΩ
    D. 20,000 Ω/V

25. Capacitors are used in _____?_____ with a resistive attenuator.
    A. Parallel
    B. Series
    C. Parallel/series
    D. Shunt

26. Capacitors with a resistive attenuator are used to
    A. Correct for stray capacitive effects at dc
    B. Correct for stray capacitive effects at low-frequencies
    C. Correct for stray capacitive effects at high-frequencies
    D. Balance out stray inductive effects at all frequencies

27. A 1-MΩ ac attenuator is sometimes used because
    A. It draws less current
    B. It is easier to compensate
    C. 1 MΩ is all that is needed for ac measurements
    D. 1 MΩ will not load the 120-V ac line

28. The simple voltage-divider input attenuator assumes the meter amplifier draws
    A. Less than 1 μA
    B. 1/1000 the total meter current
    C. Zero current
    D. Less than 1 nA

Meter Rectifier

Meter-rectifier circuits are used in electronic meters to convert ac voltages into dc voltages. The dc voltages are then used to deflect the meter. In most electronic meters the rectifier comes after the amplifier. But, as you can see, the vacuum-tube voltmeter is different from the transistorized voltmeter. Simple block diagrams of each are shown in Fig. 5-8.

Figure 5-9 shows a simplified schematic diagram of a vacuum-tube voltmeter. You can see the VTVM attenuator is split. Part comes before the rectifier input and part after the rectifier output. Voltages below 150 V go directly to the rectifier circuit. Voltages greater than 150 V are divided by either 3.33

Stray capacitance

Compensating capacitors

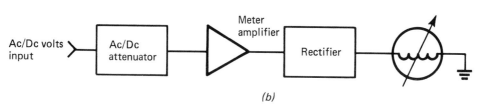

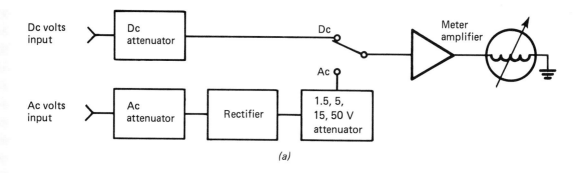

Rectifier circuits

Fig. 5-8   VTVM and TVM block diagrams.   (*a*) The **VTVM**
splits the attenuator with the rectifier circuit.   (*b*) The **TVM**
rectifies the signal after the amplifier.

or 10 before they go to the rectifier.   The dc
voltage at the rectifier output is applied to
one-half of the range switch.   This gives you
the 1.5-, 5-, 15-, 50-, and 150-V scale.   You
can see that the input impedance to the
VTVM is always 1 MΩ.

This circuit was chosen because of the
common rectifier tube used in most VTVMs.

The 6AL5 works very well over the range of 0
to 150 V.   In fact, its most linear performance
is found at the highest voltage ranges.   Frequently VTVMs have a special scale on the
meter face for the 0- to 1.5-V range.   This is
because the 6AL5 rectifier tube is not a very
linear rectifier at these very low voltages.

Two different rectifier circuits used with

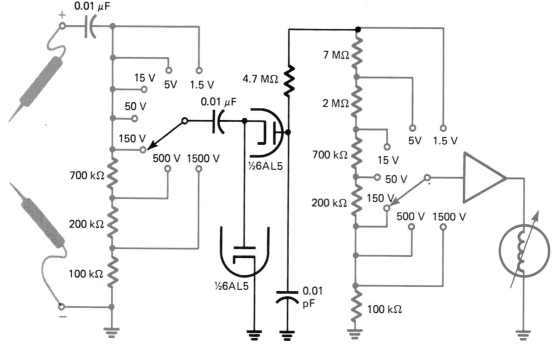

Fig. 5-9   The VTVM rectifier circuit.   Voltages up to 150 Vac
are applied directly to the 6AL5 rectifier.   Above 150 V, a
divider is used to protect the tube.

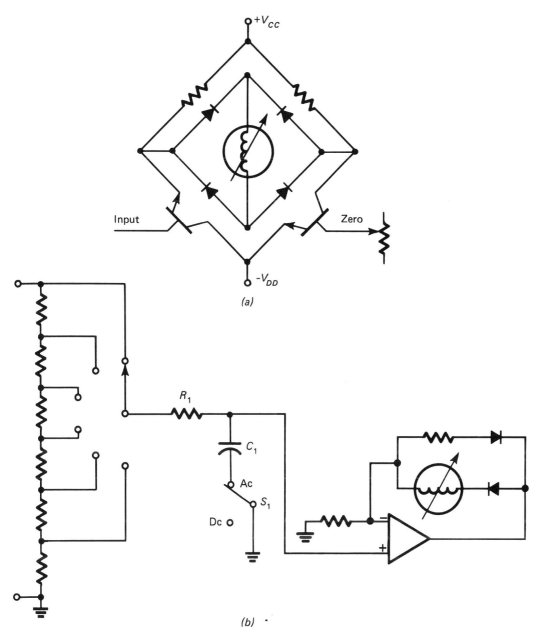

Fig. 5-10   Meter-rectifier circuits.   (a) A diode bridge in the discrete transistor amplifier.   (b) An operational rectifier is used to supply the gain and the rectifier requirements.

solid-state meters are shown in Fig. 5-10. You can see these rectifiers are part of the amplifier circuit. Therefore, different attenuators are not needed for both ac and dc measurements.

The filter in Fig. 5-10(b) is made up of resistor R_1 and capacitor C_1. This filter is switched using switch S_1. It gives filtering before the amplifier when any dc measurements are being made. This resistor-capacitor combination removes ac signals which are on the dc voltage you are trying to measure. If you

do not remove these ac signals, they will reach the meter rectifiers. When these ac signals reach the meter-rectifier circuits, they are converted to dc. They are then added to the desired dc signals. This, of course, gives incorrect readings. To prevent this, the filter is automatically switched in any time dc measurements are being made.

You will find two different meter-rectifier circuits commonly used in electronic meters. They are the peak-reading and the average-responding circuits. Both these circuits are

53

Peak-reading scales

Non-sinusoidal errors

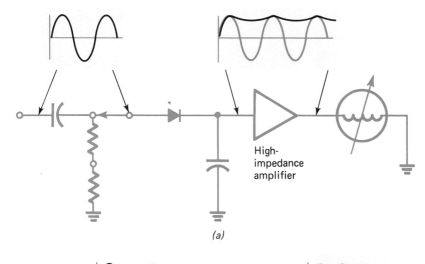

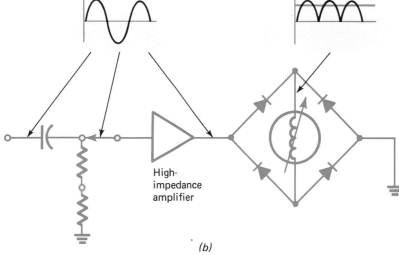

Fig. 5-11 Peak-responding versus average-responding meter circuits. (*a*) The meter is calibrated at rms = 0.707 × peak. (*b*) The meter is calibrated at rms = 1.1 × average.

calibrated to display the results in rms volts. This means neither of the two circuits actually reads rms voltages or currents. However, they are both calibrated to display rms results. Figure 5-11 shows how each circuit responds to a sine wave. Often you can tell if the meter has a peak-responding rectifier circuit because it has a peak-reading scale. Peak-reading scales are frequently quite useful. Average-reading scales are not normally included on a meter face because the user has little or no need to know the average value of a voltage or current.

Remember, the peak- and average-responding rectifiers with rms calibrated meter scales are correct only when measuring sine waves. For example, Fig. 5-12 shows what will happen if a pulse waveform is mea-

sured on the peak-responding and the average-responding circuits. A peak-responding instrument will give a different reading for 5-V pulses than the average-responding instrument. The peak-responding instrument will display.

$$V_{rms} = 0.707 \times V_{peak} = 0.707 \times 5 \text{ V} = 3.5 \text{ V}$$

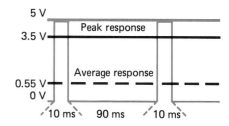

Fig. 5-12 Peak versus average response.

The average-responding instrument will respond to the 0.5-V average signal level. Therefore, the average-responding meter will display

$$V_{rms} = 1.1 \times V_{avg} = 1.1 \times 0.5 \text{ V} = 0.55 \text{ V}$$

For square waves, the rms and the average voltages are the same. Therefore for Fig. 5-12 the meter should read 0.5 V. As you can see, because neither instrument was actually displaying true rms, neither one of them displayed the correct voltage.

Self Test

29. The nonlinearity of the meter rectifier may require
   A. A different set of rectifiers
   B. A special scale for low voltages
   C. A lower meter current
   D. Getting rid of low voltage (10-V) ranges

30. If the meter rectifier is in the circuit at all times, it may
   A. Respond to ac signals while making dc measurements
   B. Be nonlinear in the low-voltage dc mode
   C. Require an operational amplifier for gain
   D. Amplify only positive signals

31. The VTVM uses a rectifier between two stages of the
   A. Meter amplifier
   B. Meter movement
   C. Dc attenuator
   D. Ac attenuator

32. If the meter rectifier comes after the meter amplifier, you do not need
   A. Ac coupling
   B. Separate ac and dc attenuators
   C. Ac filtering
   D. Dc filtering

33. The VTVM meter rectifier shown in Fig. 5-9 is
   A. Peak-responding
   B. Average-responding
   C. True-rms-responding
   D. Peak-to-peak-responding

34. The rms scales on a peak- or average-responding instrument are correct only when the input signal is a
   A. Sine wave
   B. Sine or square wave
   C. Rectangular pulse
   D. Square wave

Shunt

Meter Movement and Current Measurements

The schematic in Fig. 5-13 is one common way to make the electronic meter read current. The voltmeter function without its attenuator measures the voltage drop across a selected shunt. The shunts are selected by the current range switch.

For example, suppose an FETVOM has a maximum sensitivity of 0.25 V full scale. You can find the resistance of the 2.5-A shunt by using Ohm's law. That is, the voltage drop across the shunt must be 0.25 V when 2.5 A is flowing. Therefore, by Ohm's law,

$$R = \frac{V}{I} = \frac{0.25 \text{ V}}{2.5 \text{ }\Omega} = 0.1 \text{ }\Omega$$

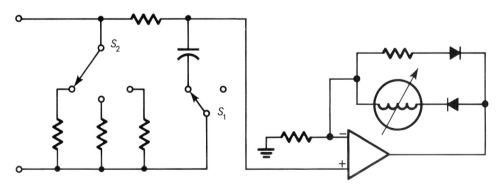

Fig. 5-13 An electronic ammeter. Switch S_2 connects the desired shunt into the current path. The meter is used at its maximum sensitivity. Both ac and dc current can be measured with this ammeter.

Using Ohm's law, we can also calculate the shunt values for the high sensitivity ranges. For example, the 1-mA shunt can be calculated for the same 0.25-V meter. The shunt value is

$$R = \frac{V}{I} = \frac{0.25 \text{ V}}{0.001 \text{ A}} = 250 \ \Omega$$

Shunt resistance error

Even though the resistance of this shunt is high (in comparison with the 2.5-A shunt), the voltage-drop is still only 0.25 V at full scale.

Shunt circuits are critical. Very low resistances found in the switch contacts, lead connections, and other places in the meter can add to the shunt resistance. If they add to the resistance, the reading will be in error. For example, suppose a poor connection added an extra 0.05 Ω to the shunt in the 2.5-A example. A full-scale meter reading is still 0.25 V, but this is now given by

$$I = \frac{V}{R} = \frac{0.25 \text{ V}}{0.1 \ \Omega + 0.05 \ \Omega} = 1.67 \text{ A}$$

In other words, 1.67 A now causes a full-scale reading on the 2.5-A scale. You can easily see that this small change in resistance has introduced about a 50% error.

$$\begin{aligned} \% \text{ error} &= \frac{\text{error}}{\text{actual value}} \times 100 \\ &= \frac{2.5 - 1.67}{1.67} \times 100 \\ &= 49.7 \end{aligned}$$

Current ranges are usually added to electronic meters when the best voltage sensitivity is quite a bit less than 1 V. Current measurement is not usually added to a meter with a 1-V sensitivity. This is because the ammeter would then cause up to 1 V drop in the circuit. Normally it is desired to keep the drop across the ammeter shunt at 0.25 V or less. For this reason, most vacuum-tube voltmeters which have a minimum full-scale sensitivity of 1.5 V do not have current-measuring capability. Like the VOM, this voltage-drop caused by putting the ammeter in the circuit, is called *insertion loss*.

As you can see from Fig. 5-13, this ammeter will measure both ac and dc currents. Ac currents were not measured with the VOM because the ammeter voltage drop would be too high. This problem is solved by adding an amplifier with plenty of gain to the meter.

We can use the gain to make ac and dc voltage or current measurements of equally high sensitivity.

Self Test

35. Calculate the shunt values required for the meter ranges in Fig. 5-14. The maximum sensitivity is 0.1 V.

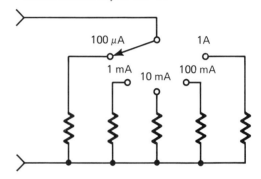

Fig. 5-14 Self test question 35.

36. How much power is dissipated in the 1-A shunt in question 35?

37. The meter in Fig. 5-13 has an *RC* filter for the amplifier to
A. Improve 60-Hz operation
B. Compensate the middle three shunt values
C. Get rid of any ac combined with the dc current
D. Eliminate the need for dc compensation

38. Meters like the VTVM do not offer current-measuring ability because of
A. The nonlinearity of the low-voltage ac range
B. The difficulty of switching shunts in a vacuum-tube circuit
C. The excessive voltage-drop required by the meter's minimum sensitivity
D. A lack of need to measure currents on early vacuum-tube circuits

39. An electronic meter has a 10,000-Ω shunt and a 0.5-V sensitivity. The ___?___ current range uses the 10,000-Ω shunt.
A. 5-μA
B. 50-μA
C. 500-μA
D. 5000-μA

Ohmmeter Circuits

Most electronic meters are built with ohmmeter circuits which are almost the same as those used in VOMs. That is, a voltmeter, a calibrating resistor, a battery, and the unknown resistance are all connected as shown in Fig. 5-15. Frequently, the ohmmeter range resistor is chosen to use 10 Ω as a midscale value. Ohmmeter ranges are selected by increasing the value of the calibration resistor by multiples of 10.

Most VOMs, VTVMs, and TVMs use a 1.5-V flashlight cell as the ohmmeter voltage source. In some solid-state electronic instruments this 1.5-V cell is replaced by an electronic power supply. Occasionally, the very high resistance ranges such as the $R \times 10,000$ or $R \times 100,000$ ranges use a 9-V battery. This is so lower-value ohmmeter range resistors may be used.

Some solid-state instruments offer a special feature called high ohms and low ohms. The use of the 1.5-V ohmmeter source causes difficulties in making measurements in solid-state circuits. The 1.5-V potential is enough to forward-bias both silicon and germanium semiconductor junctions.

For example, if you connect an ohmmeter as shown in Fig. 5-16, you will not read the true resistance in parallel with the base of the transistor. The resistance displayed is both the 100-Ω base-emitter resistor and the resistance of the base-emitter junction. To solve this problem, an ohmmeter circuit which applies only 0.085 V to the circuit being tested is often used. The 0.085-V ohmmeter becomes the "low" ohmmeter. This voltage is not enough to forward-bias either silicon or ger-

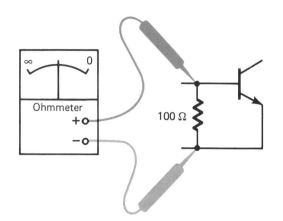

Fig. 5-16 Measuring a resistor in parallel with a PN junction.

manium junctions. When you use this ohmmeter to make the measurements shown in Fig. 5-16, it reads only the 100-Ω resistance. The base-emitter junction is not forward-biased; so it does not become a part of the meter reading.

The circuit of Fig. 5-17 is used to provide the 85-mV ohmmeter source when a 1.5-V cell is used. When the 187-Ω resistor is removed, this circuit becomes a conventional 1.5-V ohmmeter.

Normally most electronic meters have both high-ohms and low-ohms capability. This allows you to forward-bias semiconductor junctions if you choose. It also allows you not to forward-bias semiconductors if you choose not to.

The circuit of Fig. 5-18 shows an electronic constant-current power supply feeding the

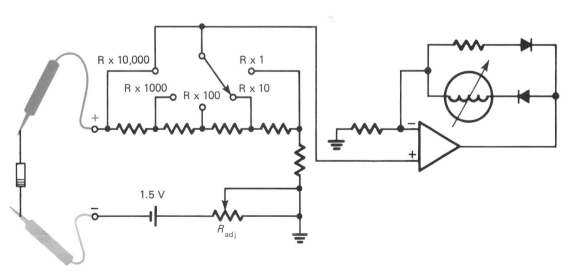

Fig. 5-15 An electronic ohmmeter. The battery may be replaced with an electronic power supply in some instruments.

57

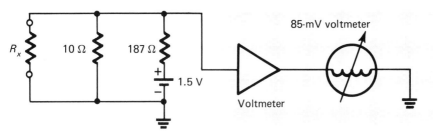

Fig. 5-17   A low-ohms circuit.   The 187-Ω resistor and the 10-Ω resistor create a voltage divider yielding a 10-Ω 0.085-V voltage source.

Reverse scale

10-Ω resistor.   If this constant-current source is set at a value of 8.5 mA, a voltage of 85 mV will be dropped across the 10-Ω resistor.   This can be seen by using Ohm's law:

$$V = IR = 0.085 \text{ A} \times 10\Omega$$
$$= 0.085 \text{ V} = 85 \text{ mV}$$

If a current of 150 mA is applied to the 10-Ω resistor, again by Ohm's law, an open-circuit voltage of 1.5 V is developed.   The circuits in Figs. 5-17 and 5-18 can be commonly found in electronic voltmeters.

The electronic constant-current source is replacing the dry-cell supply in many instruments.   The constant-current source does not need to be replaced regularly as the dry cell does.   The constant-current supply works the same way the conventional ohmmeter does.   For example, suppose the 150-mA constant-current source is driving a 10-Ω resistor.   By Ohm's law the voltage across the 10-Ω resistor is

$$V = IR = 0.15 \text{ A} \times 10 \text{ }\Omega = 1.5 \text{ V}$$

This is full scale on the meter.   It is marked ∞ ohms.   If you short the test leads, no voltage is developed across the shorted resistor.   The meter reads zero.   This point on the scale is marked 0 Ω.   If you parallel the 10-Ω resistor

with another 10-Ω resistor, the total resistance is 5 Ω.   By Ohm's law, we find the voltage drop is

$$V = IR = 0.15 \text{ A} \times 5 \text{ }\Omega = 0.75 \text{ V}$$

This is one half scale on a 1.5-V full-scale meter.   It is marked 10 Ω.

You can see that the ohmmeter operates just like the VOM ohmmeter.   The scale is simply marked in reverse.   Both circuits are in common use in electronic meters.   Fortunately, the reverse scale is a quick clue to the kind of ohmmeter voltage supply used in the particular instrument.

Self Test

40.   The low-ohms function is used when you wish to
   A.   Test field-effect transistors
   B.   Forward-bias silicon and germanium diodes
   C.   Not forward-bias semiconductor junctions
   D.   Avoid damaging FETs

41.   The electronic constant-current source is used to replace
   A.   The conventional dry cell
   B.   The calibration resistor
   C.   The ohmmeter protection circuits
   D.   The ac power supply

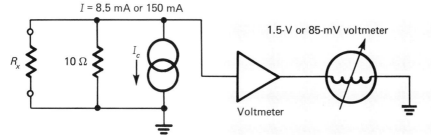

Fig. 5-18   The constant-current ohmmeter circuit.   The constant-current source I_c can be set to 0.0085 A for a 0.085-V ohmmeter or to 0.15 A for a 1.5-V ohmmeter.

Summary

1. Electronic meters give the same high input impedance on all voltage ranges. Usually this is 10 MΩ. The addition of an active amplifier allows the manufacturer to provide many special functions.

2. The electronic meter overcomes many of the VOM's limitations. It is more sensitive, and so ac current measurements ·may be made.

3. The heart of the electronic meter is either a vacuum-tube or transistor amplifier. There are a number of different forms of the electronic amplifier.

4. The meter amplifier is the basic difference between the VOM and the electronic meter. Vacuum-tube and solid-state amplifiers are both used in the bridge form. The integrated-circuit operational amplifier is also used as a meter amplifier. The meter amplifier sets the basic sensitivity of the instrument.

5. The input attenuator divides the input signal so that it fits on the instrument's most sensitive range. A simple voltage divider can be used because the meter amplifier has an almost infinite input impedance.

6. Most electronic meters use attenuators with a total resistance of 10 MΩ. VTVMs have an additional 1-MΩ resistor included in the probe tip.

The meter rectifiers are used to change ac to dc. This is done so the meter movement can show the amount of ac being measured.

7. The electronic ammeter is not much different from the passive ammeter. The voltage-drop across the shunt you select is measured with the meter amplifier/meter rectifier. Because the meter amplifier/meter rectifier works with equal sensitivity for both ac and dc measurements, you can measure both ac and dc currents.

8. The electronic ohmmeter is much like the VOM ohmmeter. Most electronic ohmmeters have a 10-Ω center scale. They use a 1.5-V source, and sometimes a 9-V source for the high resistance ranges. Some electronic meters use a 0.085-V source. This provides a low-ohm function.

Chapter Review Questions

5-1. One of the major reasons for the electronic analog meter is
(A) Its higher accuracy (B) Its greater stability (C) Its high and constant input impedance (D) Its great portability

5-2. Assume you have two 1–3–10 range sequenced instruments. One is a 1000-Ω/V VOM and the other is a 1-MΩ VTVM. Both have a maximum full scale of 3000 V. On which range does the VOM's input impedance become greater than the VTVMs?

5-3. The electronic analog meter is built with ___?___ devices.
(A) Active (B) Passive (C) Resistive (D) Capacitive

5-4. Because the electronic meter uses an amplifier, its ___?___ can be much greater than the VOM's.
(A) Accuracy (B) Sensitivity (C) Stability (D) Portability

5-5. An FETVOM often has many of the same functions as the VOM, but it also often has
(A) Low dc current (B) A very sensitive meter movement (C) A semiconductor (FET) testing capability (D) Ac current ranges

5-6. The objective of the electronic amplifier in the electronic meter is to build an instrument which
(A) Draws no current from the circuit being tested (B) Draws the same current as the VOM (FETVOMs only) (C) Draws less than the VOM's 50 μA at full scale (D) Draws less than 1 nA

5-7. Electronic analog meters usually use ___?___ for the ac/dc voltmeter input.
(A) A switched set of multiplier resistors (B) A voltage divider in a 1–10–100 sequence (C) A voltage divider in a 1–3–10 sequence (D) A compensated voltage divider

5-8. Many modern electronic meters use ____?____ meter amplifier. (A) A vacuum-tube (B) An emitter-follower (C) An integrated-circuit (D) An FET bridge

5-9. What is the purpose of the zero control on a meter amplifier?

5-10. An input attenuator is compensated to (A) Correct for 60-Hz losses (B) Provide a constant input capacitance from range to range (C) Be sure the voltage divider has the same division ratio at high frequencies as it does at dc (D) Provide ac coupling when you want to measure only the ac of an ac/dc waveform

5-11. The purpose of the meter rectifier is to convert ac signals into dc signals. Most modern electronic meters (A) Switch the rectifier into the circuit on the ac function only (B) Use a true rms meter rectifier circuit (C) Use a peak-responding rectifier with rms calibration (D) Keep the rectifier in the circuit at all times and switch in a filter to remove ac in the dc functions

5-12. Your new ACTVM has an average-responding rms calibrated meter rectifier. You know that the readings are accurate only when you are measuring (A) Ac-coupled waveforms (B) Sine waves (C) Square waves (D) Pulses

5-13. Most modern electronic meters include the ac current function because (A) Modern electronic circuits often have ac current test points (B) The meter amplifier gives enough sensitivity so the insertion loss is the same as for the dc ammeter (C) The ac shunts can be compensated easily in an electronic meter (D) Electronic meters have large scales which allow the required nonlinear presentation

5-14. The purpose of the high- and low-ohms function is to (A) Let you make resistance measurements in circuits with semiconductor devices (B) Let you make resistance measurements on high- and low-power circuits (C) Provide a method to test for low-voltage breakdowns (D) Conserve the instrument battery power

Answers to Self Tests

1. *A*
3. 500-V range, no different. They both have 10-MΩ inputs, which will load the circuits identically.

4. *A*	18. *B*	32. *B*
5. *A*	19. *A*	33. *A*
6. *C*	20. *D*	34. *A*
7. *A*	21. *A*	35. 100 μA = 1 kΩ
8. *C*	22. *A*	1 mA = 100 Ω
9. *B*	23. *C*	10 mA = 10 Ω
10. *B*	24. *B*	100 mA = 1 Ω
11. *A*	25. *A*	1 A = 0.1 Ω
12. *C*	26. *C*	36. 0.1W
13. *A*	27. *B*	37. *C*
14. *B*	28. *C*	38. *C*
15. *B*	29. *B*	39. *B*
16. 8 MΩ	30. *A*	40. *C*
17. *B*	31. *D*	41. *A*

2.

Range, volts	VOM		Electronic	
	Input *I*	Input *Z*	Input *I*	Input *Z*
1	50 μA	20 kΩ	0.1 μA	10 MΩ
2.5	50 μA	50 kΩ	0.25 μA	10 MΩ
5.0	50 μA	100 kΩ	0.5 μA	10 MΩ
10	50 μA	200 kΩ	1 μA	10 MΩ
25	50 μA	500 kΩ	2.5 μA	10 MΩ
50	50 μA	1 MΩ	5 μA	10 MΩ
100	50 μA	2 MΩ	10 μA	10 MΩ
250	50 μA	5 MΩ	25 μA	10 MΩ
500	50 μA	10 MΩ	50 μA	10 MΩ
1000	50 μA	20 MΩ	100 μA	10 MΩ

Electronic-Meter Specifications

■ This chapter discusses the specifications for analog electronic meters. It is important for you to know these specifications in order to be able to use electronic meters properly.

In this chapter you will learn the features of different types of analog electronic meters. You will also learn the common electronic-meter functions and the typical ranges for these functions. In addition, you will become familiar with accuracy specifications for electronic meters as well as other important specifications for these instruments.

6-1 INTRODUCTION

The specifications for an electronic meter are like the specifications for the VOM. They often use ac measurements and dc measurements as a dividing point. Most manufacturers give one specification for dc measurements and reduce the specification by approximately 50% for ac measurements. This reduction is to include the errors in the rectifier circuits. Accuracies of 2% dc and 3% ac are quite typical. Most manufacturers specify the meter position (horizontal or vertical) for these accuracies. Frequently the meter must be horizontal to meet its accuracy specifications. Figure 6-1 shows a typical electronic meter (an FETVOM) which has many of the specifications discussed in this chapter.

We know all instruments have limitations. Of course, the electronic meter has limitations as well. This chapter covers many of them.

6-2 FUNCTIONS AND RANGES

One of the major differences between the VOM and the electronic meter is the addition of the ac current-measurement function. Other functions and controls provide zero adjust, automatic polarity indication, high and low ohms, and other specialized features.

A number of different range sequences are used for electronic meters. The ranges 1.5, 5,

15, 50, 150, 500, and 1500 V are very common for vacuum-tube voltmeters. Normally, the voltage ranges are identical for both the ac and the dc voltage functions. TVMs tend to be somewhat more sensitive, with typical ranges being 0.1, 0.3, 1, 3, 10, 30, 100, 300, and 1000 V full scale. The 1–3–10 sequence is often chosen, as the range switch then operates in steps of 10 dB. This makes decibel readings over more than one range much easier. Having the range switch step in 10-dB increments is particularly convenient when you are measuring ac voltage. Current ranges tend to start in the 100-μA full-scale area. They frequently use decade steps to get to the 1- to 2-A full-scale region.

Self Test

1. One particular FETVOM has ac and dc voltage and current ranges as well as a resistance function. The ___?___ functions are not found on most VTVMs.
   A. Ac current and ohms
   B. Ohms and ac voltage
   C. Dc voltage and dc current
   D. Ac current and dc current

2. A new FETVOM uses the 1–3–10 range sequence. When you look closely at the meter, you see it actually has a full-scale value of 3.16 on the 3 range. This is because
   A. It is easier to do a standard calibration at 3.16 V

From page 61:
Range
sequences

1–3–10
sequence

On this page:
Accuracy

Shunt
capacitance

Fig. 6-1 The Weston 666 FETVOM. This electronic meter features high sensitivity, portability, and a constant-current ohmmeter source. Refer to this figure as you study the different specifications and features. (Courtesy of Weston Instruments Division of Sangamo Weston, Inc.)

B. A 10-dB voltage ratio is actually 3.16 to 1, not 3 to 1
C. This gives accurate peak-to-rms voltage and current conversions
D. This gives accurate average-to-rms voltage and current conversions

3. If you see an instrument with the zero adjust function you know it
   A. Is an electronic meter
   B. Is a VOM
   C. Can be either a VOM or an electronic meter
   D. Must be a VTVM

4. You are working with a new electronic voltmeter. There are two LED indicators in the meter face. One of them is marked plus and the other one is marked minus. When you see these, you suspect that the instrument is
   A. A VOM with plus and minus overload indicators
   B. An electronic meter with plus and minus overload indicators
   C. An ac voltmeter with warning indicators to tell if you are measuring dc voltages

D. An electronic meter with automatic polarity selection and indicators

6-3 INPUT IMPEDANCE

As noted earlier, most VTVMs have an 11-MΩ dc input impedance. Most other electronic meters have a 10-MΩ input impedance. The accuracy of the input impedance is usually 1% or so. The input impedance for ac ranges depends very much on the instrument you are using. Most VTVMs use a 1-MΩ ac input impedance. Many lower-cost TVMs and FETVOMs use a 1-MΩ input impedance for ac measurements. The better TVMs and FETVOMs, and ac voltmeters use a 10-MΩ input impedance.

Normally, input-impedance specifications for ac ranges will include the shunt-capacitance specification. This tells you the input capacitance you may expect at the voltmeter terminals. In most cases, a compensated attenuator is used. The capacitance then is the series value of all the capacitors used in the input attenuator.

Note: This specification does not include stray capacitance that you may introduce by connecting test leads to the instrument. This is an additional capacitance which the manufacturer cannot control and therefore does not specify. Remember, any capacitance you add further loads the ac circuits. An input capacitance between 30 and 100 picofarads (pF) is typical for an electronic meter.

An input capacitance of 100 pF can be significant, especially if the instrument uses a 10-MΩ attenuator. For example, at a frequency of 1 kHz the capacitive reactance of the 100-pF input capacitance is

$$X_C = \frac{1}{6.28\,fC} = \frac{1}{6.28 \times 1000 \times 100 \times 10^{-12}}$$
$$= 1.6 \times 10^6 \ \Omega$$
$$= 1.6 \ \text{M}\Omega$$

This 1.6-MΩ reactance is significantly lower than the 10-MΩ resistive load of the input attenuator. It will also have a substantial effect on a 1-MΩ resistive attenuator. All this simply means you must pay close attention to the amount of capacitive loading the voltmeter is placing on higher-frequency ac circuits. Remember, any loading will disturb the circuit you are measuring to some extent. The amount that the circuit is disturbed depends on the circuit impedance.

6-4 SHUNT SPECIFICATIONS

Electronic-meter shunts are usually specified in the same way VOM shunts are. A manufacturer may specify either the shunt resistance or the maximum voltage-drop caused by inserting the ammeter in the circuit. You can convert from the resistance specification to the insertion-loss specification by simply using Ohm's law. A good rule of thumb is that the maximum voltage-drop across the shunt will be the same as the most sensitive voltmeter range that the instrument has.

Self Test

5. A very good TVM specifies the ac voltmeter input impedance at 10 MΩ shunted by 30 pF. The input impedance at 1000 Hz is
   A. 5.3 MΩ capacitive
   B. 10 MΩ resistive
   C. 10 MΩ resistive shunted by 0.5 MΩ capacitive
   D. 10 MΩ resistive shunted by 5.3 MΩ capacitive

6. The VTVM uses a 1-MΩ resistor in the probe tip. This resistor with the cable capacitance forms a low-pass filter. This filter will remove ac from the dc signal you are measuring. It also makes the dc input impedance 11 MΩ. What is the current drawn from a 150-V circuit for a 10-MΩ TVM? For an 11-MΩ VTVM?

7. A particular FETVOM has a 0.25-V maximum sensitivity. The 25-mA shunt actually has a 30-pF shunt capacitance. Why doesn't the manufacturer specify this shunt capacitance for this particular instrument, which can measure current to 100 kHz?

8. If you were using the FETVOM discussed in self test question 7, you would expect the ammeter insertion loss to be
   A. 2.5 V   C. 25 mV
   B. 250 mV   D. 100 mV

6-5 ACCURACY AND FREQUENCY RESPONSE

As noted in the introduction to this section, the accuracy of the electronic meter is usually divided between the ac and the dc functions. Accuracy is normally given as plus or minus a percentage of full scale ($\pm \%$). A few very good instruments give the accuracy as plus or minus a percentage of the reading. When you see the instrument accuracy specified as plus or minus a percentage of the reading, you will be safe in assuming that it is a very high quality instrument using precision parts and an excellent meter movement. If the accuracy specification does not say whether it is a percentage of full scale or a percentage of reading, you are safe to assume that the manufacturer means that it is a percentage of full scale.

Dc Accuracy

The dc accuracy specification is quite simple. The meter amplifier is usually built and then calibrated so it introduces little or no error. The accuracy specification then tells us how good the input attenuator and the meter movement are. As noted earlier, plus or minus 2% of full-scale dc accuracy is fairly common. This usually means there is only a few tenths of a percent error in the attenuator, and the rest of the error is in the meter movement itself.

Temperature Range

On very good laboratory-grade electronic meters you may see temperature included in the error specification. This can be done in two ways. First, the instrument may be specified to operate over a given temperature range. For example, a particular instrument may be specified over the range of plus 10°C to plus 40°C. Second, a specification may indicate that the accuracy will change so many parts per million (ppm) per degree Celsius (n ppm/°C). Remember 1 ppm equals 0.0001%.

Ac Accuracy

Ac accuracy is usually specified at one single frequency. Normally, line frequency (60 Hz) is chosen for the specification. Additional error can be expected when you make measurements at frequencies other than 60 Hz. This error comes from the frequency versus gain problems in the meter amplifier and the meter rectifier. Most of the problem is in the meter-rectifier circuit itself.

Like the VOM, two different methods of specifying ac accuracy are commonly used for

Percentage of error

Decibels

Ohmmeter accuracy

Ammeter accuracy

Frequency range

electronic meters. Frequency response is the accuracy over the working frequency range for the particular instrument. It may be specified as a percentage of error or it may be specified in decibels. Sometimes the frequency-response specification will be different as you change voltage ranges. Of course, this leads to a fairly complicated ac voltmeter accuracy specification. When you are making critical ac measurements at frequencies below 20 Hz or above 1000 Hz, you should look carefully at your meter's frequency-response specifications. If the measurement is at all important, be sure to look further than a distributor's catalog or some other abbreviated data sheet. Usually the only full specifications will be given in a detail specification sheet or the instruction manual itself. Significant errors (10% or more) are not unusual when making measurements at the outside limits of the instrument's working frequency range. Also note, many VTVMs and low-cost TVMs have very poor accuracies on the most sensitive ac voltage ranges. Again, this is because of the rectifier nonlinearities which we have discussed before.

Ohmmeter Accuracy

The ohmmeter function may be given an accuracy specification. Usually this accuracy is specified as plus or minus a number of degrees of arc on the meter face. Plus or minus 3° is common.

Ammeter Accuracy

The ammeter function usually has accuracy specifications for the ac range which are the same as the accuracy specifications for the ac voltmeter function. The accuracy specifications for the dc ammeter are usually the same as the accuracy specifications for the dc voltmeter function. For example, a given electronic meter has a ±2% of full-scale dc voltmeter accuracy specification. This instrument also has a ±3% ac voltmeter accuracy. For this instrument the dc ammeter specification is ±2%, except on the 3-A dc range. Here the specification is ±3%. This extra 1% of error indicates that the manufacturer finds the low-resistance shunt difficult to build.

The ac ammeter specification reads plus or minus 3%, except for the 3-A ac range. Here

the specification reads plus or minus 4%. This expanded specification allows for the difficulty of building the shunt.

Frequency Range

The frequency range of electronic meters changes a great deal from model to model and manufacturer to manufacturer. Many early VTVMs had a working frequency range to 1 MHz. However, when the first solid-state instruments were introduced, the frequency response was often limited to 50 or 100 kHz. This reduction in frequency response was caused by the difficulty instrument manufacturers had in building wide-band solid-state amplifiers. Today we find many different frequency-response specifications. You should examine the specification carefully in relation to the job you need to do. It is too easy to get an electronic meter which just won't work for the particular high-frequency job you are doing. Again, be sure to read a detailed set of instrument specifications. Often limitations on frequency response are discussed only in the instruction manual itself. Sometimes you will find the frequency-response limitations are only for particular voltage ranges. That is to say, the low-voltage ranges will have one frequency response and the high-voltage ranges will have a different frequency response.

Self Test

9. When using an electronic meter to measure voltages in the audio range, you would expect the accuracies to be ____?____ the accuracy of power-line measurements.
   A. Better than
   B. The same as
   C. Worse than
   D. Unrelated to

10. The usable upper frequency limit of electronic meters is
   A. 1 MHz
   B. 100 kHz
   C. From 50 kHz to over 1 MHz
   D. Dependent on the dc voltmeter frequency response

11. A typical laboratory TVM is specified to have an accuracy of ±1.5% at 25°C. You will be operating this voltmeter at 32°F. In addition to the basic error specifications you find an additional specification

telling you to degrade the accuracy by 200 ppm/°C. What accuracy can you expect during this field work?

12. The usable ac frequency response of an electronic meter may be reduced to 50% or more for the highest voltage ranges. This reduction in frequency response is because of the high impedances involved in the input attenuator. Normally, you will not find this frequency-response limitation on any of the current ranges because
A. Ammeters are not usually used at high currents and high frequencies
B. The current shunts on all ranges have relatively low impedance
C. The ammeter input capacitance swamps out the high-frequency inaccuracies
D. The ammeter specification already includes the meter-rectifier frequency considerations

6-6 LINE ISOLATION

Electronic meters which are powered from the ac power-line frequency have one serious drawback. They do not have line isolation. Line isolation means the common terminal of the instrument is not connected to the case or to the ac power-line ground. A specification is normally given indicating the maximum safe voltage you can have between the instrument's common terminal and the power-line ground. If you have an instrument which does not have line isolation, you will find both the instrument case and the "negative" or "common" test-lead terminal connected to

the ground side of the ac power line. This means that you may not connect the negative lead to any part of the circuit not at ground potential. Of course, if you should connect this test lead to a part of the circuit which is not at ground potential, you will short this circuit to ground.

If, for example, you wish to make a voltage measurement between points A and B in the circuit of Fig. 6-2(a), this becomes a real problem. If you are using a VOM or an electronic meter which is isolated from the ac power line, the measurement which is shown in Fig. 6-2(b) may be used.

If you are using a voltmeter which is not line-isolated, you must use the series of measurements shown in Fig. 6-3. These measurements must then be followed by the calculation

$$V_{AB} = V_A - V_B$$

Unfortunately, this technique can lead to serious measurement errors if the voltages at points A and B are high and the difference between them is small. For example, assume the voltage at point A is 110 V and the voltage at point B is 100 V. You measure each point with a meter having 3% accuracy and a 150-V range.

The 110-V measurement is in error by ±4.5 V, as is the 100-V measurement. However, the errors at two points close to the same point on the scale and on the same range are likely to be the same. The question now is how accurately did you make these two measurements? For example, if each one was within ±2%, the measurements at each point

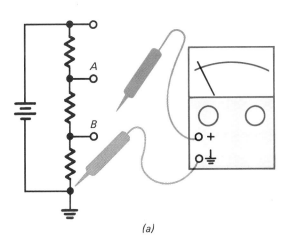

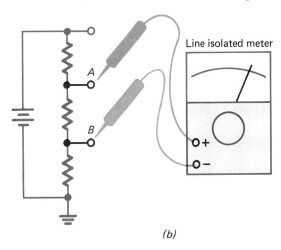

Fig. 6-2 Two different ways to measure a floating voltage.
(a) The common terminal is connected to circuit ground.
(b) A line-isolated instrument is used.

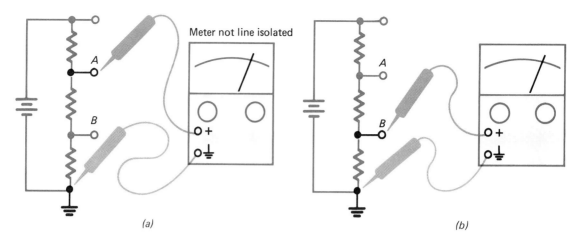

Meter not line isolated

(a)   *(b)*

Fig. 6-3   Using a non-isolated meter to measure a floating voltage.   (*a*) The voltage from **A** to ground is measured (V_A).
(*b*) The voltage from B to ground is measured (V_B).

could be 108 to 112 V and 98 to 102 V.   The
difference could therefore be 6 to 14 V, or
greater than 100% error!

6-7 AC RESPONSE

Ac voltmeter readings normally depend on
the rectifier circuit, as previously discussed.
These are converted to rms readings by a specially marked meter scale.   This assumes the
input signal is a sine wave.   If you are measuring ac signals which are not true sine
waves, the calibration of the meter is no
longer correct.   The manufacturer normally
specifies the method of ac conversion used.
This is done so that you may correct your
readings if you know the waveform of the
signal being measured.

Self Test

13. Most VTVMs do not have line isolation.
     What will happen if the common test lead
     is connected to the "hot" side of the ac
     power line?

14. A portable TVM is specified to have 500-V
     line isolation when in the ac line-operated
     mode.   The line isolation is not specified
     when the instrument is operated in the
     battery mode.   Why?

15. You are working on some logic circuits.
     You have your choice of two different
     instruments.   The first has a peak-
     responding ac voltmeter.   The second
     has an average-responding ac voltmeter.
     Which one would you choose?   Why?

Summary
1. There are two major electronic-meter
specifications.   They tell you what functions
the meter has and what the range sequence is.
2. Newer instruments have amplifiers with
gain and usually have 0.25-V full-scale maximum sensitivity or better.   The 1–3–10 range
sequence is quite popular because it means
range switching is in 10-dB steps.
3. Most electronic meters have a 10-MΩ
input impedance on the dc voltage ranges.
They may have either a 1-MΩ or a 10-MΩ
input impedance for their ac voltmeter functions.
4. Usually the ac voltmeter specification includes an input capacitance specification.
5. Shunts are specified either by stating the

maximum insertion voltage-drop or by stating
the shunt resistance for each range.
6. The accuracy specification for an electronic meter is usually given for both the ac
and dc functions.
7. The basic ac error specification is usually
given at a single frequency.   Often 60 Hz is
chosen as the reference frequency.
8. Line isolation and ac response are specified
features of an electronic meter.   The line isolation tells us if the common instrument terminal is connected to the ac power-line
ground.   The ac response specification tells
what type of meter-rectifier circuit has been
used in the design of the instrument.

Chapter Review Questions

6-1. The basic electronic-meter specifications tell you
(A) The meter's sensitivity (B) The type of active devices it uses (C) The meter's functions and ranges (D) The meter's voltage range

6-2. Typically a modern electronic meter has a basic sensitivity of _____?_____ or greater.
(A) 2.5 V (B) 1 V (C) 0.25 V (D) 0.1 V

6-3. Zero adjust, autopolarity, and high and low ohms are typically found on
(A) All electronic meters (B) Modern electronic meters (C) VOMs (D) TVMs

6-4. Typically a good modern electronic meter has
(A) A 10-MΩ input parallel with 30 to 100 pF (B) A 10-MΩ dc input and a 1-MΩ ac input (C) 20,000 Ω/V dc and 5000 Ω/V ac (D) A variable input capacitance

6-5. A VTVM often has a 1-MΩ resistor in the probe tip. The purpose of this is to
(A) Increase the input impedance (B) Keep the voltage across the other divider resistors at a lower value (C) Act as a filter, using the cable capacitance, to remove ac from the input (D) Compensate the probe

6-6. Usually the ammeter insertion loss on an electronic meter will be about the same as the instrument's
(A) Maximum voltage sensitivity (B) Meter-movement sensitivity (C) Minimum voltage sensitivity (D) Ohmmeter offset

6-7. Most electronic meters will have two basic accuracy specifications. These are
(A) The voltmeter and ammeter specifications (B) The voltmeter/ammeter and ohmmeter specifications (C) The manual and the autopolarity specifications (D) The ac and dc specifications

6-8. An electronic voltmeter is specified to have $\pm 3\%$ accuracy at 60 Hz and will be within ± 1 dB from 20 Hz to 100 kHz. This means a 100-V ac 100-kHz reading on the 100-V range could be in error by
(A) ± 3 V (B) ± 4 V (C) ± 10 V (D) ± 13 V

6-9. Why would you want line isolation on an instrument to be used in a TV repair shop?

Answers to Self Tests

1. *D*
2. *B*
3. *A*
4. *D*
5. *D*
6. 15 μA; 13.6 μA
7. The shunt will be 10 Ω. At 100 kHz, 30 pF = 53.1 kΩ. The 10-Ω shunt shorts out any capacitive effects.

8. *B*
9. *C*
10. *C*
11. $\pm 2\%$
12. *B*
13. The hot line will be shorted to the common line.
14. Because there is no con-

nection to the line in the battery mode. The isolation is the breakdown voltage of the case.

15. Peak-responding so you can see low-duty-cycle pulses which have a high peak value but a very low average value.

The Digital Meter

- This chapter discusses the digital meter. The use of digital meters in electricity and electronics is increasing.

  In this chapter you will learn the advantages of the digital meter over the analog meter. You will also become familiar with the two types of analog-to-digital converters, will be able to compare them, and will learn to draw diagrams of them.

7-1 INTRODUCTION

In the first six chapters we looked at the analog meter, the VOM, and the electronic meter. These three instruments have one common feature. On each one, you read the position of a pointer on a meter scale. This is to say, each one uses an analog readout.

The digital meter is much like the electronic analog meter. The main difference is that the analog meter movement is replaced with a digital converter. Two different types of digital converters are in use today. We will discuss the block diagrams of each of these digital meters.

The two digital meters are the single-slope and the dual-slope analog-to-digital converters. You need to know how they work before you can understand digital instruments in the next chapter.

7-2 ADVANTAGES OF THE DIGITAL METER

When you use an analog meter, the accuracy of your reading depends a lot on your ability to guess exactly where the pointer is on the meter scale. On large meters, your ability to guess the pointer position may be as good as 1%. On smaller meters, your ability to guess the pointer position is much less.

If a very good meter movement is used, a mirrored scale may be provided. Using the mirrored scale, you can make sure you are perpendicular to the meter face. This gets rid of the parallax error we discussed in Chap. 2. Using the meter with the mirrored scale, you may get accuracies of $\frac{1}{2}$% or better. Yet for many applications, $\frac{1}{2}$% accuracy may not be good enough!

The digital meter offers two improvements. First, it has far greater resolution. The $\frac{1}{2}$% analog meter lets us resolve 1 part in 200. A digital meter with a full scale of 1999 lets us resolve one part in 2000. A full scale of 19,999 gives a resolution of 1 part in 20,000! Second, the digital meter usually has accuracies close to its resolution. A 1999 digital meter may well have 0.1 to 0.05% accuracy.

Analog meters bring out another human error. They often have many scales on one meter face. This makes it easy to make another mistake. You may use the wrong scale!

The digital meter ends many of these problems. The analog meter movement is gone. In its place are electronic circuits which convert dc into a digital display. Figure 7-1 shows a digital instrument. There is no guessing about the number being displayed. It is not 1.472. It is not 1.474. It is 1.473. When you move your head from right to left, you will not change the reading as you do on an analog meter. In other words, there is no parallax error. These readings are still subject to some range mistakes. For example, the reading could be read as either 1.473 kΩ or 1.473 MΩ. However, this error is much less likely to happen than a simple wrong-scale error.

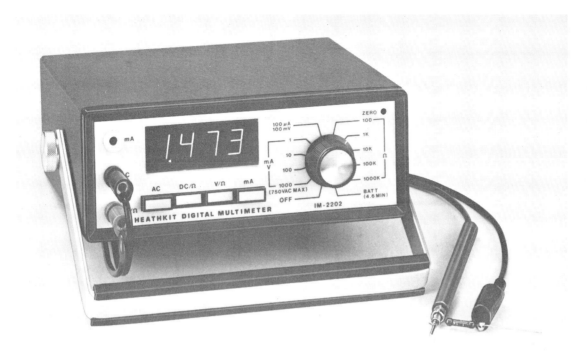

From page 68:
Resolution

Accuracy

Scale errors

No parallax
error

On this page:
Analog to
digital converter

Fig. 7-1 This digital multimeter is being used on the 1000 Ω range to read the value of a 1473-Ω resistor. Note that this 3½-digit instrument is operating properly 47.3% overranged. (Courtesy of Heath Company)

Digital meters also have another advantage. Inexperienced persons can take accurate readings with almost no training. The same people must be trained in the skill of reading an analog meter. For these reasons and others we will discuss, digital meters have replaced analog meters in many applications.

Self Test

1. A digital meter has a full scale of 199. It therefore has a maximum resolution of 1 part in 200. This is another way of saying it has ½% resolution. This meter's resolution is ____?____ a very good analog meter.
   A. Much better than
   B. Only as good as
   C. Much poorer than
   D. 100% better than

2. Which of the following is *not* a feature of the digital meter?
   A. The lack of parallax error
   B. Reading can be made by inexperienced operators
   C. A mirrored scale
   D. The lack of multiple scales

3. The digital meter ____?____ the analog meter.
   A. Has totally replaced
   B. Has partly replaced
   C. Has developed new markets, and therefore has not at all replaced
   D. Will totally replace

4. You are using a VOM with a 1–3–10 range sequence. You take a voltage reading, and record it as 0.5 V. Later a friend, using a digital meter, takes a reading and records it as 1.49 V. Why is there a difference between these two readings?

5. A digital meter has a full-scale reading of 1000. It has a resolution of
   A. 1 part in 100,000 (0.001%)
   B. 1 part in 10,000 (0.01%)
   C. 1 part in 1000 (0.1%)
   D. 1 part in 100 (1.0%)

7-3 THE ANALOG-TO-DIGITAL CONVERTER

The analog-to-digital converter is a somewhat complex circuit. It converts a voltage into the signals to drive a digital display. Two types of analog-to-digital converter are commonly

69

Single-slope converter

Dual-slope converter

Charging a capacitor

Constant-current source

used in digital meters today. These are the single-slope converter and the dual-slope converter.

The single-slope converter has a very low cost. It is less accurate and less stable than the dual-slope converter. The dual-slope converter costs more but gets rid of many of the inaccuracies found in the single-slope converter.

Most digital instruments use the dual-slope analog-to-digital converter. Many digital voltmeters (DVM) also use dual-slope analog-to-digital conversion. There are a few other analog-to-digital conversion circuits. They are used for special-purpose, high-accuracy instruments. Single-slope conversion is usually reserved for very low cost applications.

Basic Concepts

Single-slope and dual-slope converters use the same electronic-circuit fundamentals for analog-to-digital conversion. Each one charges a capacitor with a constant current. The current used to charge the capacitor is

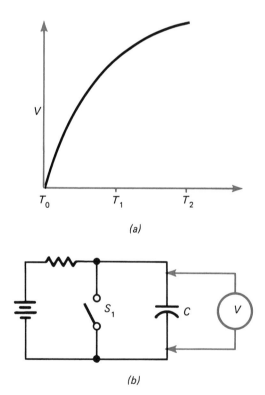

Fig. 7-2 (*a*) Nonlinear charging of a capacitor. (*b*) The voltage across the capacitor increases after S_1 in is opened.

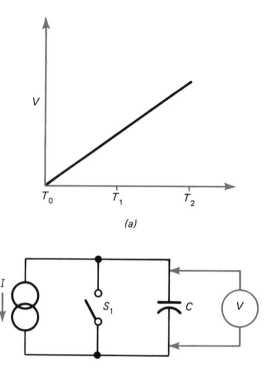

Fig. 7-3 (*a*) Linear charging of a capacitor. (*b*) The voltage across the capacitor increases linearly after S_1 is opened.

directly proportional to the unknown voltage that is being measured. Each one uses digital timing circuits. These digital timing circuits are used to display the time required to charge the capacitor. For a small unknown voltage, less charge is stored on the capacitor than for a large unknown voltage. Common digital counting and display circuits are used to time this charging and to display the results.

In order to understand how the single-slope and dual-slope converters work, you must first review the principles of charging a capacitor. Figure 7-2 shows the familiar curve of the voltage across the capacitor as it is charged through a resistor from a voltage source. As you can easily see, the charging curve is not a straight line (linear). That is to say, much more charging takes place during the first half of the charging cycle $(T_0 - T_1)$ than takes place during the second half of the charging cycle $(T_1 - T_2)$.

If, however, we charge a capacitor from a constant-current source, as shown in Fig. 7-3, we find the voltage across the capacitor increases linearly in time. That is, the voltage across the capacitor at a time (T_1) which is halfway through the charging cycle is exactly one-half the voltage across the capacitor at the end of the charging cycle (T_2). Likewise, the voltage one-quarter of the way into the charging cycle will be one-quarter of the fully charged voltage.

The amount of time needed to charge the capacitor depends upon the amount of current used. If we change the current, we change the charging time. This circuit is, therefore, used to convert a current value into a time value.

Self Test

6. The dual-slope analog-to-digital converter is ____?____ analog-to-digital converter.
   A. The most accurate
   B. For common instruments, the most popular
   C. The only analog-to-digital converter besides the single-slope
   D. By far the most accurate

7. Both the single-slope and the dual-slope analog-to-digital converters use a capacitor charged by a constant-current source. This circuit converts a voltage or current into time. A capacitor charged by a voltage source is not used because
   A. It is linear in time
   B. It is not linear in time
   C. Voltage sources are not easily controlled
   D. We only want to make current measurements

8. Figure 7-4 shows the voltage/time curve of a capacitor charged with a constant-current source. Draw a curve showing

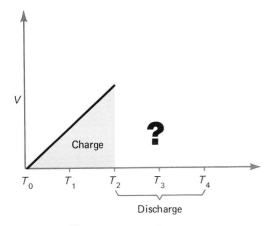

Fig. 7-4 Self test question 8.

what will happen if the capacitor is then discharged by the same constant-current value.

9. We say the voltage developed across a capacitor charged by a constant-current source increases linearly in time. This means that the time required to increase from 0 to 25% is ____?____ the time required to increase from 75 to 100%.
   A. The same as
   B. One-half
   C. Twice
   D. One-quarter
   E. Four times

The Single-Slope Converter

Charging the capacitor from a constant-current source allows us to convert an electrical quantity (current or voltage) into time. How is this used to make a digital meter?

Figure 7-5 shows the schematic for a simple single-slope converter. Two voltages are applied to the two inputs of a voltage comparator. The output of the voltage comparator is a logic 1. It stays like this while the voltage at the noninverting input (+) of the comparator is greater than the voltage at the inverting input (−) of the comparator.

In this particular circuit, a capacitor, a constant-current source, and a shorting switch are connected between the inverting input of the comparator and ground.

The noninverting input of the comparator is connected to the unknown voltage. Timing, which is directly proportional to the unknown voltage, can now take place.

For example, assume the unknown voltage is 0.5 V. When the switch (S_1) across the capacitor is open, we also start a digital timer. The capacitor has no charge at the moment the switch is open. Therefore, with no charge, there is no voltage across the capacitor. The voltage across the capacitor will begin to increase as it did in Fig. 7-3(a). At any given point in time, the capacitor voltage is

$$V = \frac{I}{C} \times T$$

Using this formula, and assuming a 10-μF capacitor is charged by a 10-μA current, we are able to find the capacitor voltage at different times. A few of these calculations are

Voltage comparator

Digital timer

Comparator
stability

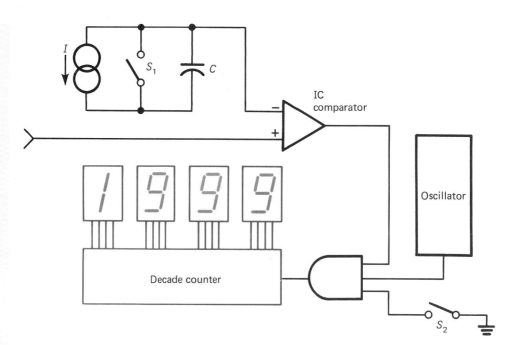

Fig. 7-5 The single-slope converter. The simplest but least accurate analog-to-digital converter.

done below.

$$V = \frac{I}{C} \times T = \frac{10 \ \mu A}{10 \ \mu F} \times 0.1 \ s = 0.1 \ V$$

$$V = \frac{I}{C} \times T = \frac{10 \ \mu A}{10 \ \mu F} \times 0.5 \ s = 0.5 \ V$$

$$V = \frac{I}{C} \times T = \frac{10 \ \mu A}{10 \ \mu F} \times 0.75 \ s = 0.75 \ V$$

$$V = \frac{I}{C} \times T = \frac{10 \ \mu A}{10 \ \mu F} \times 1 \ s = 1 \ V$$

Once the capacitor charges to 0.5 V, the output of the comparator changes. That is to say, it goes from a logic 1 to a logic 0. This happens because the voltage on the capacitor is now equal to the unknown voltage connected to the noninverting input of the comparator. When the output of the comparator changes, we stop the timing. In this particular example, 500 ms was required to charge the capacitor to a voltage equal to the unknown voltage. If the unknown voltage had been 0.75 V, 750 ms would be required.

The output of the comparator is connected to a digital timer. The timer is started when the switch (S_2) is open. It is stopped when the comparator changes its output state. The timer displays a number which can be directly read as voltage.

The digital timing circuit is shown in Fig. 7-5. It is built with a number of decade count-

ers, latches, decoder/drivers, displays, and a 1000-Hz oscillator. The 1000-Hz oscillator generates 1000 counts per second, or one count every millisecond.

In the above example the timer will total 500 counts before the comparator shuts off the counting. Of course, if the input voltage is increased to 750 mV, there will be 750 counts in the counter. In this example the time displayed can be read directly as a number of millivolts.

There are many small differences in single-slope-converter design. However, most single-slope converters basically are this simple form.

Accuracy of Single-Slope Converters

Unfortunately, the single-slope converter has a number of real problems. In order to be accurate, the single-slope converter depends on the stability of the comparator. The output of most comparators does not change when the inputs are exactly equal. A slight voltage difference between the inverting and the noninverting input terminals is required to make the change. If the required voltage difference drifts, the calibration of the single-slope converter changes.

For example, suppose the single-slope converter is calibrated when the comparator requires a 5-mV difference between the two inputs. If the comparator is heated slightly, it requires 10 mV. The additional 5 mV is error, and the 0.5-V measurement which earlier took 500 ms now takes 505 ms. This is a 1% error.

Single-slope converters have a second source of error. If the capacitance of the single-slope converter's capacitor changes, the calibration of the single-slope converter changes. If the capacitance decreases, less time is required to charge the capacitor. A 1% change in the capacitor will result in a 1% change in the time. This will cause a 1% change in the measurement.

A third source of error is the oscillator. If the oscillator frequency changes after calibration is complete, it causes an additional error. For example, suppose the single-slope converter was calibrated with an oscillator whose frequency was exactly 1000 Hz. Later, the oscillator's frequency drifts 1%. The oscillator's frequency is now 1010 Hz. When we measure the 0.5-V unknown signal, we should count 500 pulses. But the oscillator's frequency is higher than when it was calibrated. Therefore, we now count 505 pulses. This gives us a 1% error in the reading.

Unfortunately, all three of these changes can and will occur in actual comparators, capacitors, and oscillators. These problems limit the accuracy of single-slope converters to $\frac{1}{2}$% or less.

This does not mean the single-slope converter is useless. It just means it cannot be used in high-accuracy applications. This is especially true for applications which require good stability over time and temperature.

The single-slope converter is also subject to error because of input noise. For example, suppose you are measuring a 10-V dc source. Also suppose this source has a 1-V peak 60-Hz sine wave on it. At any point in time the 10-V signal can be anywhere between +11 V and +9 V. This difference is caused by the sine wave adding to and subtracting from the 10-V dc signal.

If the unknown signal is at +9 V when the single-slope converter's capacitor reaches +9 V, a −1-V error occurs. If the 60 Hz on the input signal is timed just right, it could cause an 11-V measurement to take place. That is, in the worst case, the signal does not trip the comparator until the unknown signal has reached 11 V. Of course, the comparator can trip anywhere between +9 and +11 V.

This problem can be somewhat reduced by using a resistor/capacitor filter. This filter is just like the one used for the TVM. A filter won't remove all the input noise, and some of this problem will still remain.

Self Test

10. A 1-μF capacitor is charged by a 20-μA constant-current source. This means the capacitor voltage will be charged to 20 V at the end of 1 s. At the end of 100 ms the capacitor voltage will be
    A. 100 mV
    B. 200 mV
    C. 1 V
    D. 2 V

11. A single-slope converter uses the capacitor/constant-current source in self test question 10. The comparator output changes when the capacitor charges to 4 V. The unknown voltage is
    A. 1V
    B. 2 V
    C. 4 V
    D. 8 V

12. In the single-slope converter of self test question 11, the charging time is
    A. 100 ms
    B. 200 ms
    C. 400 ms
    D. 800 ms

13. When the single-slope converter in self test question 12 is done with the measurement, the counter/display circuits show 400 counts. The frequency of the oscillator is
    A. 100 Hz
    B. 200 Hz
    C. 1000 Hz
    D. 2000 Hz

14. If the oscillator frequency in self test question 13 drifts 1% lower than the calibration frequency, the reading will change to
    A. 440 counts
    B. 404 counts
    C. 396 counts
    D. 360 counts

15. A drift in the single-slope ____?____ is not thought of as a basic source of error.
    A. Charging capacitor
    B. Unknown input voltage
    C. Oscillator frequency
    D. Comparator offset voltage

16. A single-slope converter will
   A. Measure 60-Hz noise on the unknown dc voltage exactly
   B. Read in error if a 60-Hz noise reaches the converter
   C. Always read the peak positive value of any 60-Hz noise
   D. Always read the peak negative value of any 60-Hz noise

7-4 THE DUAL-SLOPE CONVERTER

The dual-slope converter is designed to get rid of some of the single-slope converter's problems. A block diagram of a simplified dual-slope converter is shown in Fig. 7-6. The normal waveforms and timing diagrams are shown in Figs. 7-7 and 7-8. The dual-slope converter gets rid of errors due to comparator, capacitor, and oscillator drift. It does this by correcting for new values each time a measurement is made. This correction is performed on one of the "slopes." The measurement is performed on the other slope.

The operational amplifier in Fig. 7-6 is connected as an integrator. The voltage across the capacitor is exactly the same as the output voltage of the operational amplifier. The capacitor in an integrator is charged by a constant current. The charging current is given by

$$I = \frac{V}{R}$$

where V is the voltage on resistor R. As you can see from the schematic, V can be either the reference voltage or the unknown input voltage. The switch S_1 selects which voltage is applied to the resistor.

This all means the capacitor C is charged with a constant current. The charging current is directly proportional to the voltage at the input to the resistor R.

At the start (T_0) of the measurement cycle, the integrator is connected to the unknown voltage. At this time (T_0), the output of the integrator is usually slightly below 0 V. That is to say, the integrating capacitor has a slightly negative charge. When the measurement cycle starts, the integrator output voltage becomes more and more positive. Soon the comparator output changes from a logic 0 to a logic 1. This happens as the integrator's output voltage passes through 0 V toward a more positive value. This time is labeled T_1 on the waveforms. As soon as the comparator output changes, the counters begin to count pulses from the oscillator.

The counting continues until the counter overflows. This happens at T_2. For example, if the counter is a divide-by-2000 counter, it contains 2000 counts. The overflow signal is received on the 2001st pulse.

As soon as the counter overflows, the input to the integrator is switched from the unknown voltage to a reference voltage of the opposite polarity. The polarity of the referenced voltage is opposite to the polarity of the

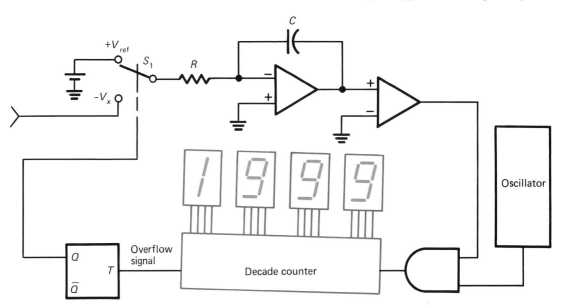

Fig. 7-6 The dual-slope converter. This is the most common analog-to-digital converter used in instruments of all types.

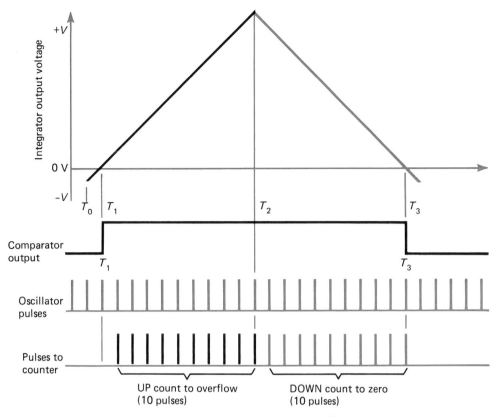

Fig. 7-7   The dual-slope waveforms for an input voltage equal to
the reference voltage.

unknown voltage. This makes the output of the integrator reverse direction. It now starts to return to 0 V. The comparator output goes through 0 V at time T_3 on the waveform chart. The counter measures the time needed for the integrator output to pass through the comparator's zero crossing point. This is the time span T_2 to T_3 on the waveform charts.

The time span from T_2 to T_3 depends on the voltage the integrator was charged to. The discharge is done by a constant current. The constant current is produced by the reference voltage on the input resistance R.

A few examples will help illustrate this point. As you read these examples, refer to the waveforms shown in Figs. 7-7 and 7-8. These are quite simple. Only 10 counts of up ramp are used. This makes it easier to see just what is happening. Suppose, for example, that the unknown voltage is exactly 1 V. The waveforms in Fig. 7-7 show what happens. The resistor/capacitor combination of the integrator is chosen so that 1 V at the input causes the output to rise to +10 V. This happens in a time span of 1 s. The unknown voltage is +1 V. The counters will count 10

pulses during the 1-s time span. These pulses come from the 100-Hz oscillator. The tenth pulse will cause an overflow. The overflow is detected and causes the input to switch from the unknown voltage to the reference voltage. The reference voltage is also 1 V. It takes exactly 1 s to discharge the capacitor. This means that 10 counts are displayed when the counter is stopped. In this example we see the length of time required for the two slopes is exactly equal. This is because the unknown input voltage and the reference voltage were exactly equal.

Now suppose the input is connected to an unknown voltage which is one-half the reference voltage. That is to say, the input is 0.5 V. The waveforms in Fig. 7-8 show what happens. The integrator output will go to only +5 V during the 1-s initial measurement period. The input is still integrated for 10 pulses, because this time is controlled by the counter overflow. On overflow, the integrator is connected to the reference voltage. The output of the integrator now drops from 5 to 0 V. But this takes only $\frac{1}{2}$ s. This is because the 1-V reference will discharge the integrator at the rate of 10 V/s. The counter

75

Voltage ratio

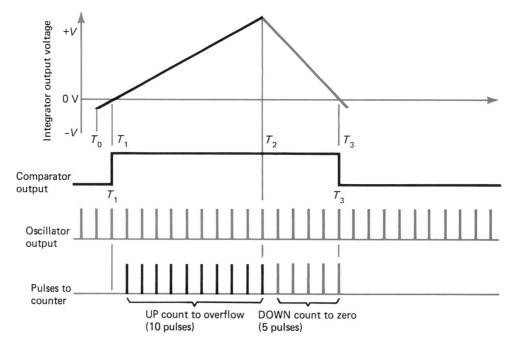

Fig. 7-8 The dual-slope waveforms for an input voltage equal to
one-half the reference voltage.

now has 5 counts. This is shown on the display. You read this as 0.5 V.

Selt Test

17. The dual-slope converter gets its name from its
    A. Two constant-current sources
    B. Use of two comparators
    C. Charge/discharge diagram
    D. Use of dual ranges

18. The discharge slope of a dual-slope converter is always at a constant rate. This is because the integrator is connected to the reference voltage during this time. Because the rate is constant, the time to discharge the capacitor from 5 V to 0 V is _____?_____ as is needed to discharge the capacitor from 10 V to 0 V.
    A. The same
    B. Twice as long
    C. One-half as long
    D. Five times as long
    E. One-fifth as long

19. The first slope on a dual-slope converter always
    A. Covers the same amount of time
    B. Is the same time as the second slope
    C. Is one-half the time of the second slope
    D. Is twice the time of the second slope

20. In the circuit of Fig. 7-6 pulses flow into the counter
    A. Any time the integrator output voltage is above 0 V
    B. Any time the integrator output is below 0 V
    C. Only during the second slope
    D. Only during the first slope

7-5 GETTING RID OF ERRORS

How does the dual-slope converter get rid of the errors we found in the single-slope converter? The dual-slope converter gets rid of the single-slope-converter errors because it simply displays the ratio of the reference voltage to the unknown voltage.

If we look carefully at the dual-slope-converter waveforms, we see that

$$\frac{V_{in}}{V_{ref}} = \frac{T_2 - T_3}{T_1 - T_2}$$

This equation tells us the measurement is a simple ratio. Comparator voltage, capacitor value, and oscillator frequency are not part of this equation. This means we can change their values, within reasonable limits, without changing the measurement at all.

We know that the comparator will not

switch at exactly 0 V. But what happens if the voltage changes after the dual-slope converter is calibrated? For example, you calibrate your DVM when the comparator has a +5-mV offset. Later you use this DVM when the comparator's offset has drifted to +10 mV. Looking back at the dual-slope-converter block diagram, we see the pulses do not go to the counters until the output of the integrator reaches +10 mV. The pulses stop flowing to the counters when the integrator's output voltage gets back to +10 mV. Therefore, the integrator now goes up to 10.01 V with a 1-V input. Before it would go up to only 10.005 V. Notice, however, the actual integrator output still ramps up by only 10.000 V while the pulses are actually being counted. Therefore, there is no change in the converter's reading because of this change in offset.

From this explanation you can see that a fast drift can cause an error. For example, if the comparator's offset was +10 mV at the start of a measurement cycle, and it was +25 mV at the end of this same measurement cycle, the ramp would be 15 mV short. Such a drift as this, in a short time, is very unlikely.

The dual-slope-converter calibration is also stable with capacitance changes. For example, we know a larger capacitor will take more charge before it has the same voltage as the smaller capacitor. If the capacitor changes its value after calibration, the maximum integration voltage changes. But we are not really interested in the integrator's output voltage. We are interested in filling up the capacitor with an unknown amount of charge, and then discharging the capacitor at a known rate. Changing the capacitor's value does not change the ratio of the charge to discharge time. Again, we must assume the capacitor is stable over one single measurement cycle. This is a reasonable assumption.

Oscillator Stability

The dual-slope converter accuracy does not depend on the long-term stability of the oscillator. The dual-slope converter requires only that the oscillator is stable during one measurement cycle.

For example, assume the oscillator was accidentally set at 1200 Hz rather than the 1000 Hz which it was when the converter was calibrated. A 1-V signal is connected to the unknown input for the time required to total 1000 pulses. This now takes 833.3 ms, not 1000 ms, because the oscillator now has a higher frequency. The integrator does not charge up to 10 V. It now can charge only to 8.333 V, because the time is shorter. But the 1-V reference is now connected to discharge the capacitor. It has to discharge it for only 833.3 ms rather than 1000 ms. In 833.3 ms, 1000 pulses of the 1200-Hz clock are counted. The voltage is displayed as 1000 mV (1.000 V).

We can see, however, that the reference voltage must be stable over time. The reference is used only on the second cycle. If the reference voltage changes, the discharge rate changes. This does not cancel out in the measurement and therefore will be a cause of error.

Noise Rejection

The dual-slope converter can also have a very good built-in 50/60-Hz noise filter. This filter is designed using the first-slope integration time.

For example, suppose the first-slope integration time is made to be exactly 100 ms. If there is 60-Hz noise on the input signal, we will integrate exactly 6 cycles. This is shown in Fig. 7-9. The 60-Hz noise adds nothing to the total charge on the capacitor at the end of 100 ms. The extra charge added by the posi-

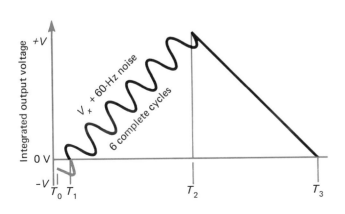

Fig. 7-9 Line-frequency noise rejection. Any line-frequency noise integrated with the dc input signal will cancel itself out because of the equal energy in the positive and negative half cycles.

50-Hz noise

Normal-mode
rejection ratio
(NMRR)

tive half cycles of the noise is exactly canceled by the extra charge subtracted by the negative half cycles of the noise.

If there is 50-Hz noise on the input signal, the integration time is for exactly 5 cycles. The same addition and subtraction process takes place. Again, the 50-Hz noise adds nothing to the total charge on the integrator.

The ability of the dual-slope converter to reject ac power-line frequency noise can be very great. Often a digital meter can measure a dc voltage with an equal amount of line-frequency noise voltage riding on it. That is, it can reject 100% noise.

This rejection does depend on the integration time being an exact multiple of the noise frequency. This means the oscillator must be stable. This is because the rejection takes place on only one slope. In other words, we did not get something for nothing. High noise rejection was gained at the expense of a high-stability oscillator.

Note: Not all dual-slope converter designs have this feature. A high (greater than 40 dB) normal-mode rejection ratio (NMRR) at line frequency only usually indicates this type of rejection in the dual-slope converter. The term normal-mode rejection ratio indicates the ability of the instrument to reject noise riding on the input signal.

Self Test

21. The dual-slope converter gets rid of errors from capacitor, comparator, and oscillator drifts by
    A. Using them in both slopes
    B. Using them in the analog and digital circuits
    C. Matching capacitor drifts to oscillator drifts so they cancel
    D. Integrating for exactly 100 ms

22. A change in the comparator offset voltage changes
    A. The reference voltage
    B. Both the start and the stop voltage
    C. The time the comparator output is at a logic 1
    D. The displayed count at the end of a measurement cycle

23. If the capacitor value drifts after a digital meter is calibrated the energy stored in the capacitor during a measurement
    A. Increases by exactly the same amount
    B. Decreases by exactly the same amount
    C. Is inversely proportional to the change
    D. Does not change; just the charge voltage changes

24. The dual-slope converter accuracy does not depend on the stability of its oscillator. However, the _____?_____ of the dual-slope converter may depend a great deal on its oscillator stability.
    A. Counting accuracy
    B. Reference voltage
    C. Line-frequency noise rejection
    D. Display update

25. The accuracy of the dual-slope converter does depend on the _____?_____ stability over time.
    A. Integrator resistor/capacitor circuit
    B. Comparator offset drift
    C. Reference voltage
    D. Oscillator

26. A very high line-frequency noise rejection may come from
    A. A simple *RC* filter in the input circuit
    B. Making the first-slope integration time the same time as an exact multiple of the line frequency
    C. Operating the instrument on a battery supply
    D. Making the first slope exactly 110 ms long

Summary

1. The digital meter converts an unknown signal into numbers instead of a pointer position as the analog meter does.

2. The person using the analog meter must be trained to read it. Much less training is needed for reading the digital meter.

3. The digital meter also has much higher resolution and accuracy than the analog meter.

4. The analog-to-digital converter is the electronic circuit which replaces the magnetic coil and spring on an analog meter. The digital displays replace the pointer and scale.

5. Two analog-to-digital converters are in common use today: the single-slope converter and the dual-slope converter. The dual-slope converter costs more than the single-slope converter but is much more accurate.

6. The single-slope converter measures the time to charge a capacitor to be equal to the unknown voltage. A digital display shows the amount of time required to charge the capacitor.

7. The single-slope converter uses a few parts. Its accuracy depends on the stability of these parts.

8. The dual-slope converter is designed to get rid of many of the single-slope converter's errors. It does this by comparing the unknown voltage to a reference voltage.

9. The dual-slope converter gets rid of three major single-slope problems: comparator offset drift, capacitor drift, and oscillator drift.

10. A number of different dual-slope analog-to-digital converter designs are in use today. Some of these designs feature "auto-zeroing," which removes zero drifts in the integrator itself.

11. The auto-zero designs have a special measurement cycle which checks to see that a zero input is displayed as a zero output. If the converter does not show zero output for a zero input, a correction is put in to make the output zero.

12. Some dual-slope converter designs have a very high line-frequency noise-rejection (NMRR) capability. This is done by making the time for the first slope an exact multiple of the line frequency.

Chapter Review Questions

7-1. A digital meter has a full scale of 1999. This meter has an accuracy of 0.1% and a resolution of
(A) 0.5% (B) 0.1% (C) 0.05% (D) 0.01%

7-2. A digital meter eliminates errors caused by
(A) Parallax (B) A vibrating pointer (C) Multiple scales (D) Unskilled operators (E) All of these reasons

7-3. It is often said the digital meter will never replace the analog meter for making "trend" measurements. Why?

7-4. The single-slope converter is ____?____ analog-to-digital converter.
(A) The most accurate (B) The most popular for common instruments (C) The lowest-cost (D) The most expensive

7-5. Both the dual-slope and the single-slope analog-to-digital converter use a capacitor charged by a constant-current source to convert the unknown voltage to
(A) Current (B) Voltage (C) Time (D) Resistance

7-6. A capacitor is charged by a constant current because the voltage across the capacitor
(A) Rises linearly in time (B) Is inversely proportional to time (C) Is exponential in time (D) Is independent of the value of the capacitor

7-7. A single-slope converter will not read in error if the ____?____ drifts.
(A) Reference voltage (B) Capacitor sizes (C) Comparator (D) Input voltage

7-8. A single-slope converter measures the time needed to charge a capacitor to ____?____. This time is measured digitally and displayed as a voltage.
(A) One-fourth the unknown value (B) One-half the unknown value (C) Three-fourths the unknown value (D) The unknown value

7-9. A dual-slope analog-to-digital converter will read in error if the ____?____ drifts.

(A) Reference voltage (B) Capacitor size (C) Oscillator frequency (D) Comparator

7-10. The amount of charge on the dual-slope capacitor after the first slope depends on
(A) The reference voltage (B) The discharge time (C) The unknown input voltage (D) The comparator drift

7-11. The discharge rate on the second slope
(A) Depends on the unknown voltage (B) Is always the same (C) Always takes the same amount of time (D) Is used to open the comparator

7-12. The dual-slope converter makes a ratio measurement comparing the unknown voltage to the reference voltage. In doing this it
(A) Cancels comparator drift (B) Cancels oscillator drift (C) Cancels capacitor drift (D) Does all of the above

7-13. Even though the dual-slope converter may not need a high-stability ____?____ for accuracy, it may be required to give a high normal-mode rejection ratio.
(A) Capacitor (B) Oscillator (C) Comparator (D) Reference

7-14. Your new DVM has a high NMRR from 20 Hz to 10 kHz. From this specification you know the instrument has
(A) A fast slope integration time which is an exact multiple of the line period (B) Battery operation (C) A high-stability reference voltage (D) An input filter

Answers to Self Tests

1. *B*
2. *C*
3. *B*
4. Because you misread the VOM. You used the 0–1-V scale instead of the 0–3-V scale.
5. *C*
6. *B*
7. *B*
8. The voltage will pass through zero at T_4.

9. *A*	18. *C*
10. *D*	19. *A*
11. *C*	20. *A*
12. *B*	21. *A*
13. *D*	22. *B*
14. *C*	23. *D*
15. *B*	24. *C*
16. *B*	25. *C*
17. *C*	26. *B*

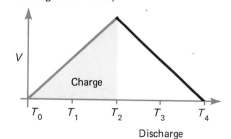

Digital Instruments

- This chapter discusses digital instruments, including the digital voltmeter, the digital multimeter, the digital volt-ohmmeter, and the digital panel meter.

  In this chapter, you will learn to identify and describe each of these instruments, including basic features and approximate accuracies. You will also learn how to draw a block diagram for a digital multimeter.

8-1 INTRODUCTION

There are several digital instruments that are built using the digital converter discussed in Chap. 7. Some of these instruments, such as the digital panel meter, are quite simple. Others are complex electronic meters with digital displays. Often these instruments have many special features which are not found on analog instruments.

We will look briefly at each of the major different digital instruments. We will then look at the digital multimeter in detail. Finally, we will review the specifications and features of the digital multimeter. The digital multimeter is chosen because it has all the features found in any of the less complex instruments.

8-2 DIGITAL METERS

In many ways the digital meter may be thought of as a replacement for the analog meter except that it uses a digital display. The basic digital meter normally responds to voltage. A typical digital meter might have a range of 0 to 200 mV. Once we have a digital meter, we can construct a number of test instruments. We just connect the right amplifiers, attenuators, shunts, ohmmeter circuits, etc.

Quite a variety of digital instruments are available. Each of these goes by a different name. Each one of them is just a little bit different from the others. We will now look at some of the more common instruments using a digital display.

The Digital Voltmeter

The digital voltmeter (DVM) was the earliest digital meter. A few DVMs were actually designed during the days of vacuum-tube circuits. When speaking of DVMs, most manufacturers are talking about instruments which will only measure voltage. Some will measure both ac and dc voltages. A few also have resistance functions. A simplified diagram of the typical dc-only DVM is shown in Fig. 8-1. True DVMs usually have very high accuracy (0.05 to 0.005%).

The Digital Multimeter

The digital multimeter (DMM) is a name given to the instrument which includes voltage, current, and resistance functions. Most DMMs are power-line-operated and are moderately high accuracy instruments. They usually have a fundamental dc accuracy in the area of 0.1 to 0.05%. To show the digital multimeter properly, we add shunts, the ohmmeter, and the ac rectifier to the diagram shown in Fig. 8-1. The new block diagram is shown in Fig. 8-2. Figure 8-3 shows a typical DMM.

The Digital Volt-Ohmmeter

The digital volt-ohmmeter (DVOM) is the name used to describe a portable DMM. As its name implies, the DVOM has all the functions of a volt-ohmmeter. The DVOM accuracies are normally in the 0.5% area. DVOMs almost always operate from an in-

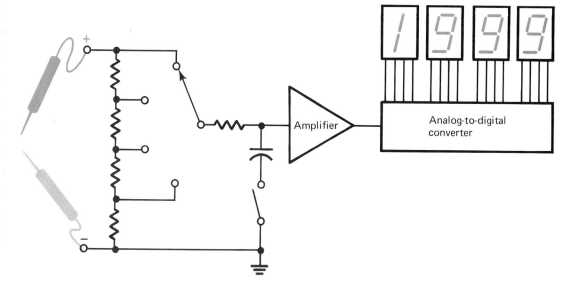

Fig. 8-1   The DVM.   This is the simplest of the digital meters.
The analog-to-digital converter and the digital displays replace
the analog meter movement used in the electronic analog meter.

ternal battery supply.   The block diagram of
the DVOM is the same as that of the DMM.
The major differences between the DVOM
and the DMM are portability and accuracy.

The Digital Panel Meter

The digital panel meter (DPM) is the digital
equivalent of the analog panel meter.   DPMs
are designed to replace analog panel meters in
many industrial applications.   DPMs vary a
great deal in accuracy, number of digits, and
input voltage ranges.   Almost all DPMs have
a single input range.   This input range is set
by the manufacturer before the DPM is purchased.   Figure 8-4 shows a typical DPM.
Figure 8-5 is an inside view of this DPM.
From this photograph you can easily see the
electronics required to replace the simple analog panel meter.

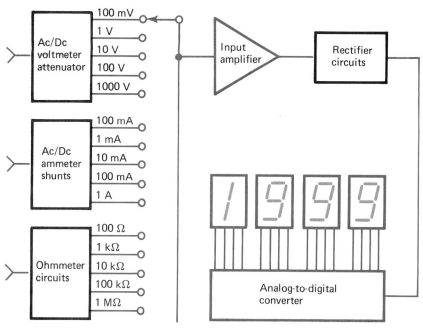

Fig. 8-2   The DMM.   The DMM has full measurement capability.   Current, resistance, and ac circuits are added to the DVM
of Fig. 8-1.

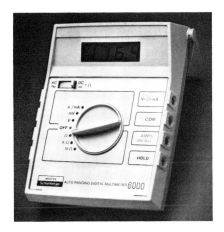

Fig. 8-3 An auto-ranging digital multimeter. This is a typical low-cost high-performance digital replacement for the VOM. (Courtesy Weston Instruments Division of Sangamo Weston, Inc.)

You must be sure the proper names are used when speaking of a digital instrument. Many persons confuse the DVM with the DMM and the DVOM. To them, they are all DVMs, even if they have many other functions in addition to the voltmeter function.

Self Test

1. The DVOM is named after the VOM. This is because it has the same functions and is
   A. Highly accurate   C. Line-operated
   B. Portable   D. Lightweight

2. The DVM typically has accuracy specifications in the range of
   A. 1 to 0.01%   C. 0.05 to 0.005%
   B. 0.5 to 0.05%   D. 0.005 to 0.0005%

3. The digital panel meter is used to
   A. Replace the analog panel meter
   B. Construct a digital volt-ohmmeter
   C. Give high (0.05 to 0.005%) accuracies
   D. Offer multirange capability in a panel instrument

Fig. 8-4 The DPM. This is the digital replacement for the analog panel meter. This instrument has many uses once filled exclusively by the analog panel meters. (Courtesy of Weston Instruments Division of Sangamo Weston, Inc.)

Fig. 8-5 Inside the DPM. These electronic circuits are needed to replace the analog panel meter. Notice the transformer for power and the two large integrated circuits which contain most of the circuitry. (Courtesy of Weston Instruments Division of Sangamo Weston, Inc.)

4. The ___?___ is usually built using low power logic, displays, and analog circuits because it is often battery-operated.
   A. DVOM   C. DMM
   B. DVM   D. DPM

5. You would think that the DVM was the first digital instrument because
   A. Early work needed only voltage measurements
   B. Early work needed voltage, current, and resistance measurements

   C. Early instruments were complicated enough with just the voltage function
   D. Current measurement requires the use of the IC operational amplifier

8-3 A DMM BLOCK DIAGRAM

In the block diagram of a typical digital multimeter (see Fig. 8-2), we see many of the circuits which we have used in analog multimeters. In this section we will see there is really little difference between the analog and the digital meter. DVM, DPMs, DVOMs, etc., can be made by just reworking the diagram to get rid of the unnecessary parts. We will look only at the differences between the digital and the analog meter. You should always remember the accuracy of the components of a digital instrument will be much greater than the accuracy of the same components in an analog instrument. This is because the digital meter itself is much more accurate than the analog meter. This puts a much greater demand on the accuracy of all the other parts of the instrument.

83

Function-
selection
switch

Input attenuator

Automatic-
ranging

Amplification

Impedance
buffering

Operational
rectifier

Accuracy

Peak-responding

Average-
responding

Linear
ohmmeter

Input Circuits

The DMM input circuits are made up of the input attenuator and the function-selection switch. The function selection allows the DMM to measure voltage, current, or resistance. The input attenuator gives the DMM its wide range of measurement. Sometimes the input attenuator is automated. The automatic-ranging feature selects the range which is best for the measurement you are making. The user is still required to select voltage, current, or resistance functions.

Amplifiers

Most digital meters do not have enough sensitivity for a basic digital multimeter. Therefore, some amplification is usually necessary before the signal goes to the analog-to-digital converter. The amplifier also provides impedance buffering to avoid loading the input attenuator. Sometimes these amplifiers change gain electronically. This is done to simplify the input attenuator design. For all practical purposes the input amplifier is just like the one used in the analog meter.

The Rectifier

Most digital meters use an operational rectifier to convert ac signals to dc. The operational rectifier does the best possible job without going to a true rms converter. The true rms converter is not used because it is very expensive. The rectifier is a real weak point in a digital meter. It is hard to build a rectifier which will have much better than 0.5% accuracy. This looks quite bad when the dc accuracy is 0.1%, 0.05%, or better! Unfortunately, the accuracy is often reduced if the input signal is not a pure sine wave. Most sources are not pure sine waves. To actually get 0.5% accuracy, the sine wave will probably have to come from a very good audio-oscillator. Ac power lines, for example, are not close enough to a pure 60-Hz sine wave to give more than 1 to 2% accuracy. There are two forms of operational rectifier. First is the operational rectifier designed to be peak-responding and rms-calibrated. Second is the operational rectifier designed to be average-responding and rms-calibrated. Obviously, both of these rectifier circuits display their results as rms voltages. However, the real measurement takes place as either a peak reading or an average reading.

Automatic Polarity

Another feature is automatic polarity selection. As you can see, the dual-slope converter we have been discussing works only with a negative input voltage. Often you want to measure both positive and negative voltages. The automatic polarity converter detects both positive and negative inputs. It then switches the converter to measure the correct input. The input can be either positive or negative. Automatic polarity dual-slope converters use two voltage references. One is used for positive measurements and the other for negative measurements. Once the polarity of the input signal is determined, it is displayed. This circuit also connects the correct reference-charge source for the down ramp. (See Fig. 8-6.)

The Ohmmeter Circuits

Ohmmeters for digital meters have a special problem we do not find with analog meters. When we used the analog meter to measure resistance, a special nonlinear meter scale was used. It is very difficult to build a nonlinear digital meter scale. This means we must build a linear ohmmeter. The linear ohmmeter is normally built as shown in Fig. 8-7.

In this circuit a constant current is supplied to the unknown resistance. A voltage is developed across the unknown resistance because of the constant current flowing through it. It is, by Ohm's law ($V = IR$), simply the product of the constant current I and the unknown resistance R. If the resistance is cut in half, the voltage across the resistance will be cut in half. This makes a linear ohmmeter.

For example, a constant current of 1 mA is chosen. The unknown resistance is 1000 Ω; so we can see 1 V is developed across the resistor. One volt is displayed as 1000 on the digital meter. This is read as 1000 Ω when the meter is in the ohmmeter function.

The voltage developed by the ohmmeter's circuit depends on the current supplied and the maximum voltmeter sensitivity. To select different ohmmeter ranges, different currents are used. For example, a meter with 1 V sensitivity uses 1 mA for the 0- to 1-kΩ range. The same meter uses a 1-μA source for the 0- to 1-MΩ range.

Some DMMs offer high and low ohmmeter ranges like the ones found on analog elec-

Fig. 8-6   The auto polarity feature determines the polarity of the input voltage and connect the correct reference voltage charge source for the down ramp.   The polarity is also indicated by the appropriate LED display.

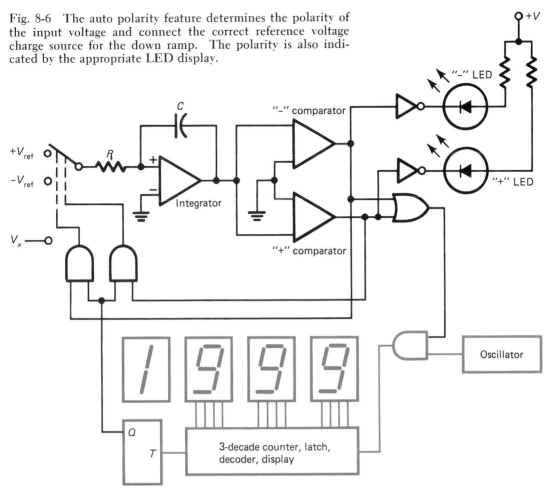

High- and low-ohms functions

tronic meters.   To do this, we use a lower test current and a higher-sensitivity voltmeter range.   For example, the 0- to 1-kΩ range in the above example used a 1-mA current and the DMM's 1-V range.   If we use the 100 mV range, we use a 100-μA constant current.   By Ohm's law, we see

$$V = I \times R = 100 \times 10^{-6}\,\text{A} \times 100\Omega$$
$$= 100 \times 10^{-3}\,\text{V} = 100\,\text{mV}$$

This provides the low-ohms function.   Remember, we must change the decimal point to give the correct reading.   We also must be sure the ohmmeter output voltage stays at least 100 mV or less.   If the upper voltage is not limited to 100 mV or less, semiconductor junctions will be forward-biased.   This will defeat the purpose of the low-ohms test position.

The Analog-to-Digital Converter and Displays

The analog-to-digital converter is made up of the integrator, the comparator, and the

counting circuits.   The display circuits show the digital output of the analog-to-digital converter.   Frequently the display circuits use latches.   This is done so the display changes only after each measurement cycle.   This way you do not see the pulses being counted.   All

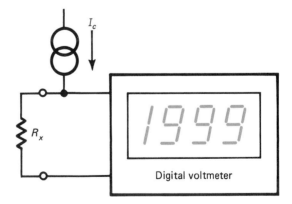

Fig. 8-7   The DMM's ohmmeter.   In this circuit the DMM's voltmeter is used to read the voltage across the unknown resistor R_x.   The voltage drop across the resistor is caused by a constant-current I_c.

85

you see is a change when the measurement changes between two measurement cycles.

Self Test

6. If you purchased an expensive DVM with an ac accuracy specified at 0.1% or better, you would suspect you had bought
   A. A true rms converter
   B. An operational rectifier
   C. A peak-reading circuit calibrated in rms
   D. An average-reading circuit calibrated in rms

7. The digital-meter range sequence is 1–10–100. This is because
   A. The input attenuators are simpler
   B. There is plenty of accuracy with decade ranging
   C. Decade ranging gives convenient 20-dB
   D. You can not have two different (0-3, 0-10) scales on a digital meter

8. The digital ohmmeter circuit uses a constant-current source driving the unknown resistor. This is used because
   A. This gives better accuracy
   B. You cannot rate a digital ohmmeter accuracy in ± degrees of arc
   C. Digital meters cannot have a nonlinear scale
   D. The analog meter uses a 10-Ω center scale

9. The block diagrams of the digital meter and the analog meter are very much alike. The two areas of difference are in the meter movement and the
   A. Accuracy of the circuit blocks
   B. Degree of portability
   C. Accuracy of ac measurements
   D. Range sequence

8-4 TYPICAL DMM SPECIFICATIONS

Generally speaking, the digital multimeter is a more complex instrument than the analog multimeter. Most of this added complexity comes from the much higher accuracies you find in the digital instrument. Therefore, it can have additional types of errors. This means there is normally a more detailed set of specifications for the meter being used.

The digital meter usually has two differences in the error specifications. First, most digital meters give the error specification as a percentage of the reading. You will re-

member that the analog meter used a percentage of full scale. This can mean quite an improvement for less than full-scale reading. Second, most digital meters follow the accuracy specification by the statement "± 1 digit." This means the meter accuracy includes a ± 1-digit error at all points on the scale. At the full scale this does not mean much. At lower-scale readings this can be a substantial error. We will review a number of these specifications, and some typical values will be given.

Ranges

As with the analog meter, the digital voltmeter specifies the ranges over which it will operate. Ranges for the digital voltmeter are usually given in decade steps. That is, they are multiples of 10. For example, the ranges for a digital multimeter might be 200 mV, 2 V, 20 V, 200 V, and 2000 V full scale. Often the ranges are given as multiples of 2 as in the above example. Many manufacturers call this 100% overranging capability. They consider, for example, the 200-V range to be a 100-V range with 100-V (100%) overrange capability. No interdecade ranges like the 1–3–10 or 1.5–5–15 ranges are chosen. This is because the digits naturally display readings in decade multiples. The current and resistance ranges are also in decade steps.

Full scale is usually specified as 199 or 1999. The 199 instrument is often called a 2½-digit instrument. The 1999 instrument is often called a 3½-digit instrument. The ½ digit is there because of the 100% overrange capability. The 1999 specification is used because often the 2000 reading will flash, blank out, or otherwise indicate an overrange condition exists. In other words, it often will not be displayed. The maximum reading will be 1999. This gives a full 2000 counts if you start counting at 0000. Some of these displays are shown in Fig. 8-8.

Accuracy

The accuracy specifications given to the voltage, current, and resistance ranges for a digital multimeter are usually much more detailed than the accuracy specifications for similar ranges on an analog electronic voltmeter. For example, dc accuracies of 0.1 to 0.05% are very common. However, the specifications

Percentage of reading

± 1 digit

Decade steps

100% overranging

Full scale

½ digit

2 $\frac{1}{2}$ digits

3 $\frac{1}{2}$ digits

4 $\frac{1}{2}$ digits

Fig. 8-8 Maximum digital meter displays. Instruments whose maximum displays are 199, 1999, and 19999 are called, respectively, $2\frac{1}{2}$, $3\frac{1}{2}$, and $4\frac{1}{2}$ digit instruments.

may be given at a particular temperature (normally 25°C). In addition to the simple accuracy specification, a statement of temperature coefficient is also given.

Temperature coefficient indicates the error which you can expect for every degree of temperature change away from the specified value. A typical temperature coefficient might be 200 parts per million per degree Celsius (0.02%/°C). Often an analog voltmeter may have the same temperature coefficient. However, its resolution, that is, its ability to detect small changes, is nowhere near as great as that of the digital meter. Therefore, the specification is not usually used.

Sometimes the manufacturer may tell you that the instrument will work over a temperature range. For example, one typical voltmeter might operate over the range of 15 to 30°C. The manufacturer may give a temperature coefficient for use outside that temperature range. In other cases no temperature coefficients may be given at all.

The ac accuracy specifications are nearly always much worse than the dc accuracy specifications. The ac accuracy specifications must take into account the rectifier circuits. They also must recognize that the rectifier circuit is either average- or peak-responding. This average or peak output from the rectifier is converted to an rms equivalent before it is displayed. Very few waveforms are exact sine waves. This means the conversion from average or peak to rms will be inaccurate. Ac accuracies which lie between 0.5 and 1% are quite common on DMMs with 0.1 to 0.05% dc accuracy.

Most ac meter specifications also include the operating frequency range for the particular instrument. The operating frequency range may depend on the voltage range being used. It is usually the same frequency range for all current ranges. In many cases the frequency range is quite limited. An upper limit of 10 kHz is not unusual. A lower limit of 20 Hz is common.

Dc current range accuracies are normally very much like the dc voltage accuracies. However, a reduction in accuracy may occur on the high current ranges, because it is difficult to obtain accurate low-resistance shunts.

The ac current ranges normally include rectifier-error specifications as well as shunt accuracy. Both ac and dc will include temperature-coefficient specifications.

As the resistance readings are now linear, error specification is quite possible. Digital multimeters normally specify the resistance reading errors at about the same accuracy as the digital voltmeter specification for the same instrument. There is often an exception for an extremely high resistance range. For example, the 20-MΩ full-scale resistance range may have a different error specification. This range may have a reduced error specification because of the extremely low currents used in the ohmmeter circuits. These currents not only cause a voltage-drop across the unknown resistor but must also charge any stray capacitances around the test leads and across the resistor being measured. Although these capacitances are small, they require some length of time to charge when very low currents are involved. Therefore, a time specification may be added to the very high value ranges.

When using the digital meter, you must be very careful not to confuse accuracy and resolution. For example, you are using a 1999 meter with a 0.1% accuracy. The 0.1% accuracy specification tells you the reading is good to 1 part in 1000. The 1999 (2000 count) display gives you a resolution of 1 part in 2000. In this case you can resolve changes which are a smaller percentage than the accuracy of the instrument.

As a further example, the ac accuracy for this same instrument is specified at 0.5%. This can be read as an accuracy of 1 part in 200. In this particular case, the instrument is really only good to 199. The 1999 display makes the resolution ten times as good as the accuracy will permit! In most cases this added resolution is of no use. In fact, it often contributes to improper readings.

Normal-mode rejection ratio (NMRR)

Normal-mode filter

High line-frequency NMRR

Measuring NMRR

Number of digits

Digit size

Display type

Polarity selection

Overrange indicators

Normal-Mode Rejection Ratio

Most digital meters will give a normal-mode rejection ratio (NMRR). This specification tells you how much ac noise will be rejected by the dc voltmeter or ammeter. A normal-mode voltage is an interfering signal which is in series with the desired signal. This is shown in Fig. 8-9.

Normal-mode rejection can be from a filter or it can be from a dual-slope converter which has special timing. If a filter is used, the normal-mode rejection ratio will be given for a broad band of frequencies, such as 20 Hz to 20 kHz. A normal-mode rejection ratio specification which is limited to only the line frequency is usually an indication that the instrument has special dual-slope timing.

Normal-mode rejection ratio is usually specified in decibels above the least significant digit. For example, suppose you have a DVM with 2.000 V full scale. Another way to say this is that this range has 1 mV resolution. A 60-dB NMRR means a 1-V noise signal will change this reading by only 1 mV. One volt is one thousand times greater than 1 mV. Another way to say this is that 1 V is 60 dB greater than 1 mV. Normal-mode rejection ratio is found by increasing the input signal in series with the instrument and measuring the value that causes the least significant digit to just begin to change.

8-5 TYPICAL DMM FEATURES

As we have noted before, the digital meter is a fairly complex electronic package. This means it is often built with many operator-convenience features. Many of these features are quite useful. Many will be found only on instruments built by one particular manufacturer. Some of these features are listed below. You probably will not find the different features on one instrument.

1. *Number of digits:* Gives the instrument's accuracy and resolution. Instruments with $2\frac{1}{2}$ and $3\frac{1}{2}$ digits are the most common while ones with $4\frac{1}{2}$ or $5\frac{1}{2}$ digits are found only on very expensive DVMs.

2. *Digit size:* Digit size ranges from less than 0.1 inch high on very compact instruments to over $\frac{1}{2}$ inch high on some bench models. Digit size on digital panel meters is very important if you are going to be making readings from a distance.

3. *Display type:* There are many different types of display. LEDs and LCDs are quite popular. Vacuum fluorescent and gas-discharge displays are also very commonly used. Each display has a different advantage depending on the type of light you are working in. For example, sunlight may wash out an LED display and liquid-crystal displays do not show up in the dark. The LCD display also requires less power than a comparable LED display.

4. *Polarity selection and indication:* The dc polarity of a digital instrument may be manually switched on low-cost instruments. On most medium- and high-priced instruments, automatic polarity is used. An instrument with automatic polarity usually has indicators to show which polarity signal is being measured. Some instruments have both + and − indicators; other instruments indicate only negative measurements.

5. *Overrange indicators:* The overrange indicator tells the user when the measurement is off scale. Many different overrange indicators are used. It can be a flashing "over" or a turning off of all the

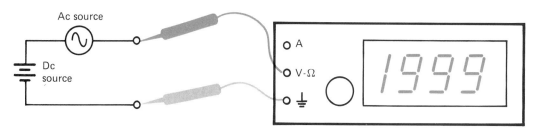

Fig. 8-9 The normal-mode signal. The ac signal source in series with the desired dc source is called a normal mode or a "series" mode signal.

digits, or sometimes just a very wrong number is displayed.

6. *Display rate:* It tells you how fast the measurement is being made. Usually three to five measurements per second are made. If the measurements are made much faster than five per second, the eye cannot follow the changes in the digits.

7. *Auto range:* An auto-ranging instrument will select the best range for the measurement. Usually the instrument will change ranges at a different voltage for increasing voltage than it will for decreasing voltage. Few instruments can auto-range the high current ranges.

8. *Input protection:* The input-protection specification indicates a maximum voltage which will not damage the instrument on any range. It usually applies to voltage, current, and resistance functions.

9. *Power:* The power source indicates whether the instrument will operate from ac only or ac and dc. Usually dc operation means it carries an internal battery pack. If battery operation is available, an operating time is usually specified before the batteries are discharged. If the batteries are rechargeable, a recharging time is usually given.

10. *Line isolation:* This specification indicates how many volts can be applied between the instrument's common terminal and the power-line ground. Usually this is limited to about 500 V dc.

11. *Size/weight:* These specifications tell you just how portable the instrument really is. Portability will be indicated by other features important to you for your situation.

Remember, each of these features can be a very important specification for a particular use. As digital instruments become more complicated, you must read the specifications more carefully. What may be a good instrument for one particular application may be very poor for another application.

Display rate

Auto range

Input protection

Power

Line isolation

Size/weight

Self Test

10. You are using a $3\frac{1}{2}$-digit (1999) DMM. It has 0.5% accuracy. The display indicates a change in the input from 1.955 to 1.950 V. You can safely assume the signal changed 5 mV. But you don't know the absolute value of the final reading any better than
    A. ± 1 V   C. ± 10 mV
    B. ± 100 mV   D. ± 1 mV

11. If you are using a DMM with a 1999 display to measure the ac line voltage, you can expect a reading accuracy of about
    A. $\pm 10\%$ C. $\pm 0.01\%$
    B. $\pm 1\%$ D. $\pm 0.001\%$

12. Your DMM has a 0.5% dc voltage accuracy. You would expect it to have ___?___ accuracy on the ohmmeter ranges.
    A. $\pm 5\%$ C. $\pm 0.05\%$
    B. $\pm 0.5\%$ D. $\pm 0.005\%$

13. You are using your VOM and DVOM to check your car's battery voltage on a $-5°$F winter day. DVOM has a -200 ppm/°C temperature coefficient. Which one will give the most realistic reading? Why?

14. Which type of instrument would you expect to use LCD (liquid-crystal displays), the DVM or the DVOM? Why?

15. You have a DVOM with a 199 display. It is specified at 1% of reading accuracy ± 1 digit. The absolute accuracy when reading 1 V is
    A. ± 40 mV   C. ± 10 mV
    B. ± 20 mV   D. ± 1 mV

Summary

1. Once we have the digital replacement for the analog meter, we can build many instruments.

2. The replacement for the simple analog panel meter is called the DPM (digital panel meter).

3. DMM stands for digital multimeter. The DMM is a line-operated instrument with ac/dc voltage and current functions as well as a resistance function.

4. The DVOM has the same functions as a DMM, but it is a less accurate portable instrument.

5. The DVM is a high-accuracy digital voltmeter.

6. The block diagram of a digital multimeter looks very much like the block diagram for an electronic analog multimeter. Both use an input attenuator, amplification, rectification, ohmmeter circuits, and a meter movement.

7. The digital meter has much higher accuracies and resolution than the analog meter. This means the specifications must be much more detailed.

Chapter Review Questions

8-1. What are the full names of the following instruments?
(*a*) DPM (*b*) DVM (*c*) DMM (*d*) DVOM

8-2. The DVOM is usually a low-power portable instrument. In functions and features it is otherwise much like the
(A) DMM (B) DPM (C) DVM (D) VOM

8-3. Typically DMMs have accuracy specifications in the range of
(A) 1 to 0.1% (B) 0.5 to 0.05% (C) 0.05 to 0.005% (D) 0.005 to 0.0005%

8-4. The block diagrams of a good analog electronic meter and a digital multimeter with the same functions will show the biggest differences in their
(A) Input attenuator (B) Meter rectifier (C) Meter movements (D) Shunts

8-5. The accuracy of most common (average- or peak-responding) meter rectifiers limits DMM ac voltmeter/ammeter accuracies to the range of
(A) 1 to 0.1% (B) 0.5 to 0.05% (C) 0.05 to 0.005% (D) 0.005 to 0.0005%

8-6. The digital ohmmeter must produce a ____?____ response to changes in the unknown resistance.
(A) Nonlinear (B) Linear (C) Inversely proportional (D) Exponential

8-7. The range sequence on all digital meters is
(A) 1–3–10 (B) 1.5–5.0–15 (C) 1–10–100 (D) 0, 10, 20 dB, etc.

8-8. Your new DMM is advertised as a $3\frac{1}{2}$-digit instrument. This means its full-scale value is
(A) 100 (B) 199 (C) 1000 (D) 1999

8-9. You are using a 0.1% of reading 1999 DVM on the 10-V range to measure a 5 V dc supply. The reading is 5±
(A) 0.02 V (B) 0.006 V (C) 0.005 V (D) 0.002 V

8-10. What would be the major disadvantage with a $2\frac{1}{2}$-digit DVOM which has only 10% overrange capability?

8-11. For most DVOMs, DMMs, etc., the accuracy of the ohmmeter function is basically the same as the voltmeter accuracy. Why?

Answers to Self Tests

1. *B*
2. *C*
3. *A*
4. *A*
5. *C*
6. *A*
7. *D*
8. *C*
9. *A*
10. *C*
11. *B*
12. *B*
13. Both will give realistic readings even though the DVOM has a −0.9% change because of temperature. This is not enough to affect a car's battery readings. *Note:* The real question is will the DVOM work at −5°F?
14. DVOM. Because LCDs are low-power displays well suited to a low-power instrument.
15. *B*

The Electronic Counter

- This chapter describes the electronic counter and its various functions. Use of the electronic counter is increasing steadily.

  In this chapter, you will learn how to identify and describe frequency, period, multiple period, time interval, and events. You will also become familiar with each major component of the electronic counter. In addition, you will learn how to diagram and explain the operation of an events counter, a frequency meter, a period meter, and a time-interval meter.

9-1 INTRODUCTION

The electric and electronic circuits you work with use many digital circuits. The electronic counter therefore is an important tool. In addition to measuring pulse and digital circuits, electronic counters do many other jobs. However, the need to work on digital circuits makes them much more useful. Electronic counters are also in much greater demand because they are constructed with low-cost digital integrated circuits. The price of an electronic counter in the mid-1960s was thousands of dollars. Within several years a multifunction electronic counter may be available for a few hundred dollars. Nearly every electronic repair shop has at least one counter. The electronic counter shown in Fig. 9-1 is a typical low-cost instrument.

The names "counter" and "frequency meter" are often confused. The digital frequency meter is a special-purpose electronic counter. It only measures and displays the frequency of an unknown signal. On the

Fig. 9-1 A low-cost electronic counter. This instrument has many of the functions typical of the integrated-circuit counters. Counters like this cost thousands of dollars in the 1960s. (Courtesy of Heath Company)

other hand, the electronic counter (usually called just a "counter") is a multifunction instrument. It makes digital frequency measurements, counts the number of pulses over a user-selected period of time, makes time-interval measurements, and measures the period of a signal.

In order to understand what the electronic counter does, we will review and define the terms frequency, period, and time interval. We will also examine the basic block diagrams of an electronic counter. Each block will be looked at to determine exactly what it does. We will review the different ways these blocks can be connected to build each of the electronic counter's instrument functions. We will also look at the input circuits needed to connect (interface) the analog world with the digital circuits inside the electronic counter.

9-2 FREQUENCY, PERIOD, AND TIME INTERVAL

As noted earlier, you must understand the terms frequency, period, and time interval before you use an electronic counter. To review these terms, we will use the waveform diagrams in Fig. 9-2.

Frequency is shown in Fig. 9-2(a). Simply, frequency is the number of cycles per second (Hz). In Fig. 9-2(a) we see the frequency of this waveform is 10 Hz. That is, 10 complete cycles of the waveform happen in a time interval of 1 s. Note that the waveform shown is not a sine wave. It does not have to be a sine wave. All we are interested in is the number of complete cycles. It does not matter how complex the waveform is.

The period of the same waveform is diagramed in Fig. 9-2(b). The period is the time required for one complete cycle to happen. You can start the time measurement at any point on the waveform. You must stop the time measurement at exactly the same point on the waveform. This is shown in greater detail in the second section of Fig. 9-2(b). Here, you can see the exact start point does not matter. You must choose the same stop point. Period and frequency are related by

$$T = \frac{1}{f}$$

where T is equal to the period in seconds and f is equal to the frequency in hertz. This rela-

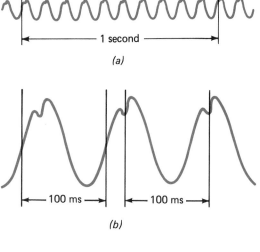

(a)

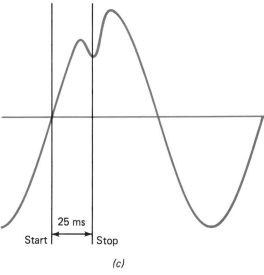

(b)

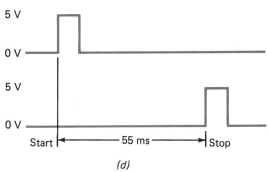

(c)

(d)

Fig. 9-2 Frequency, period, and time interval. (a) A 10-Hz waveform. That is, there are 10 cycles in the period of 1 s. (b) The $\frac{1}{10}$-s period of a 10-Hz waveform. (c) A time-interval measurement between two points on the same waveform. (d) A time-interval measurement using separate start and stop waveforms.

tionship is demonstrated in Fig. 9-2. For example, we know by counting complete cycles that the waveform in Fig. 9-2(a) has a frequency of 10 Hz. Therefore, its period should be

$$T = \frac{1}{f} = \frac{1}{10 \text{ Hz}} = 0.1 \text{ s} = 100 \text{ ms}$$

In Fig. 9-2(b) we are looking at an exploded view of one cycle of this waveform. Here we see the waveform does indeed have a period of 100 ms. Some common frequency/period relationships are: *time interval* and *period measurements*.

Frequency	Period
1 Hz	1.000 s
30 Hz	33.333 ms
60 Hz	16.667 ms
100 Hz	10.000 ms
1000 Hz	1.000 ms
15,750 Hz	63.492 μs
1 MHz	1.000 μs
10 MHz	100 ns
100 MHz	10 ns
1 GHz	1 ns

Time interval is simply the measurement of the time between two points you select. The two points can be on the same waveform, as shown in Fig. 9-2(c). They can also be two points on different start and stop waveforms. This is shown in Fig. 9-2(d). Period measurement is just a special version of time-interval measurement. That is, the start and the stop points are exactly the same. They are also on the same waveform.

Self Test

1. You are at an auto race track. You count the number of cars which pass your position during some time interval. You are making a(n) ____?____ measurement.
   A. Frequency
   B. Period
   C. Time-interval
   D. Event

2. You measure the length of time between car 56 and car 13. You are making a(n) ____?____ measurement.
   A. Frequency
   B. Period
   C. Time-interval
   D. Event

3. The yellow light goes on and all cars must hold exactly the same position for a number of laps. You measure from the time the front bumper of car 18 passes the telephone pole in front of your seat until it passes the same point again. You are making a(n) ____?____ measurement.
   A. Frequency
   B. Period
   C. Time-interval
   D. Events

4. The National Bureau of Standards' station WWV transmits on exactly 5 MHz. The period of this signal is
   A. 500 ns
   B. 50 ns
   C. 5 ns
   D. 200 ns
   E. 20 ns
   F. 2 ns

5. Often the time-interval function is used to measure pulse width. The leading edge of a positive-going pulse is used to start the time-interval movement. The ____?____ is used to stop the time-interval measurement.
   A. Next leading edge of the same pulse
   B. Next leading edge of another circuit's pulse
   C. Next trailing edge of the same pulse
   D. Next trailing edge of another circuit's pulse

6. A square wave has a frequency of 500 Hz. This means there are ____?____ positive rectangular pulses per second.
   A. 1000
   B. 500
   C. 250
   D. 100

7. The modern low-cost electronic counter is possible because of
   A. The modern low-cost digital voltmeter
   B. The need to set radio transmitters exactly on frequency
   C. The ease of making both frequency and period measurements
   D. Modern low-cost digital integrated circuits

9-3 THE ELECTRONIC COUNTER'S BUILDING BLOCKS

You can easily understand the modern electronic counter if you study the five building blocks which are used to make this instrument. You can connect these building blocks

Time interval

together in different ways to produce the different instrument functions. Even though each of these building blocks is connected to the other building blocks in a different way, the operation of each building block stays the same. Therefore, we can study each of the blocks to learn how they work. We can then combine them to build the different instrument functions. Four of the five building blocks are made up almost entirely of digital circuits. The fifth building block is the analog interface. That is, it contains the circuits which convert the signals from the outside world into signals which will drive the digital integrated circuits.

The five building blocks for an electronic counter are:

1. Decade counting and display circuits
2. Time-base oscillator
3. Divider chain
4. Gate and control circuits
5. Input amplifier and shaping circuits

Decade Counting and Display Circuits

As their name implies, the heart of the electronic counter is the circuits which count electronic pulses and display the results. Figure 9-3 is a simplified schematic diagram of a dec-

ade counting and display unit. This particular diagram uses four decades of counting and display. In addition to the decade counters, an overrange circuit is also included. The basic building block of the decade counting and display unit is a BCD (binary-coded decimal) decade counter.

One decade counter is used for each decade of counting you want to do. For example, if you want to count from zero to one thousand pulses, you must use three decade counters. If you wish to count from zero to one million pulses, you must use six decade counters.

The outputs of each decade counter are stored in a four-bit latch. The output of the four-bit latch is connected to the decoder/driver circuits for that decade. The decoder/driver circuits drive the display. Each four-bit latch stores the information contained in its decade counter at the moment it gets a memory signal pulse. The latch is used so the display shows a steady number even while the decade counter is changing.

The decoder/driver converts the BCD information from the latch into the signals used to drive the display. In most modern counters a BCD to seven-segment decoder is used. This is because seven-segment displays are normally used. The driver portion of the decoder/driver converts the logic signals into

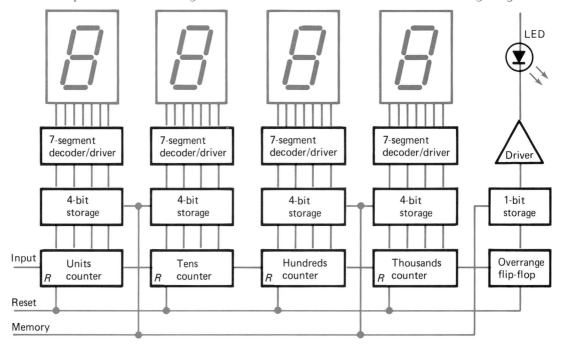

Fig. 9-3 The decade counting and display unit. This four-digit decade counting and display unit will count up to 9999 pulses. The 10,000th pulse sets the overrange flip-flop.

higher-power signals which have enough current and voltage to run the display.

In Fig. 9-3, the least significant digit is on the left-hand side of the schematic. The digit on the right-hand side of the schematic is the most significant digit. Following the most significant digit is an overrange circuit. The overrange circuit is triggered when the most significant digit goes from a nine to a zero. This circuit stores the overrange information until all the decade counters as well as the overrange circuit are reset.

The decade counting and display unit may be made up of individual integrated circuits. Each decade of counting, storage, and display may be one single integrated circuit, or the entire counting, storage, and decoder block may be one large-scale integrated circuit. It does not really matter what construction is used; the function is still the same. The decade counting and display unit counts electrical pulses supplied to its input. It then, on command, displays the total number of pulses counted. The counting period starts when the reset pulse clears the decade counters to all zeros. Each input pulse is then counted by the decade counters. When a pulse is received on the memory line, the number of counts is stored in the latches. This causes the number of counts between the reset pulse and the memory pulse to be displayed.

The counting and display circuits have some maximum electrical capabilities. These limit what the instrument can do. First, the highest frequency of the counting and display unit depends on the first decade counter's frequency limit. Second, the total number of pulses which can be counted before an overrange happens depends on the number of decade counters you used.

For example, think of a decade counting and display unit made up of six decade counters. The first decade counter is a 74S290 IC. Using six decades of counting, we can display any number of input pulses between zero and 999,999. The one-millionth pulse will overrange the decade counting and display unit.

The 74S290 has an 80-MHz capability. Therefore, the maximum rate at which you can count pulses is eighty million pulses per second. The frequency of the pulses coming from the output of the first decade counter will be one-tenth the input frequency. This means any decade counter with at least 8 MHz

or greater capability will work as the second decade counter.

Self Test

8. A digital frequency meter has eight digits. This means it has eight decade counters. The instrument's maximum frequency is 250 MHz. A frequency capability of _____?_____ is needed for the second decade counter.
   A. 8 MHz
   B. 25 MHz
   C. 80 MHz
   D. 250 MHz

9. The frequency meter in self test question 8 has eight digits. This means it can count up to _____?_____ before it overranges.
   A. 10,000,000
   B. 19,999,999
   C. 99,999,999
   D. 100,000,000

10. The decade counting and display unit has _____?_____ decade counter(s) for each digit.
    A. One   C. Seven
    B. Four   D. Ten

11. The decade counting and display unit uses latches. The latches store one set of counts
    A. To prevent a reset
    B. For high-frequency measurements
    C. While the next number is counting up
    D. To keep the memory pulse from resetting the counter

12. The purpose of the decoder/driver circuit is to
    A. Buffer the counters
    B. Convert BCD to seven-segment
    C. Drive liquid-crystal displays
    D. Convert the counter outputs into a logical and electrical signal which will drive the display

13. The overrange circuit responds
    A. Each time the most significant digit changes from nine to zero
    B. Each time the least significant digit changes from nine to zero
    C. Once in a counting cycle when the most significant digit changes from a nine to a zero
    D. Once in a counting cycle when the least significant digit changes from a nine to a zero

Overrange circuit

Counting period

Memory pulse

Upper frequency

Standard
frequency

Quartz-crystal
Oscillator

TCXO

The Time-Base Oscillator

Most of the measurements which you will make with the electronic counter compare a standard frequency (or time) to an unknown frequency (or time). Just how these comparisons are made is discussed later.

Because you are comparing an unknown frequency (or time) to a standard frequency (or time), the electronic counter must have its own frequency (or time) standard. We know that frequency and time are related by

$$f = \frac{1}{T}$$

where f is the signal's frequency and T is the signal's period. This simple formula tells us that if we have a standard frequency, we have a standard time. It also tells us that if we have a standard time, we have a standard frequency.

The most common frequency/time standard in a modern electronic counter is a quartz-crystal oscillator. A quartz-crystal oscillator is chosen because it is highly accurate. It is also very stable over a long period of time and with wide changes in temperature. The quartz-crystal oscillator is easily calibrated and relatively inexpensive to build.

Three different quartz-crystal oscillators are in use today: the simple quartz-crystal oscillator, the TCXO, and the ovened oscillator. Each of these has a different accuracy. As the oscillator becomes more accurate, it also becomes more expensive.

The simplest quartz-crystal oscillator is shown in Fig. 9-4. With this simple oscillator, a moderately stable frequency is generated. Typically, the quartz crystal is cut to operate at 1, 4, or 10 MHz. A 1-MHz crystal, for example, generates a standard frequency of 1 MHz, which has a standard time of 1 μs.

The frequency of a simple crystal oscillator changes slightly with changes in temperature. In some measurements, even these small changes are too much. The next step in improving the frequency stability, and therefore the accuracy, is to replace the simple crystal oscillator with a temperature-compensated crystal oscillator (TCXO). TCXOs make a ten times improvement in the frequency stability of the time base.

For example, the frequency of a simple crystal oscillator varies by 10 Hz when the temperature changes from 10 to 40°C. The TXCO, however, limits this change to 1 Hz for the same temperature change.

TXCOs may be built in a number of different ways. The simplified schematic in Fig. 9-5 shows a typical TXCO. In this circuit, the frequency of the crystal oscillator is adjusted by the capacitance changes from the varicap diode (VD). The voltage across the varicap diode controls the varicap's capacitance. Changes in this voltage are caused by resistance changes in the voltage-divider network, which contains thermistors (T_1 and T_2).

The voltage across the varicap controls the varicap's capacitance. The varicap capacitance controls the oscillator's frequency. Because the varicap's voltage is dependent on

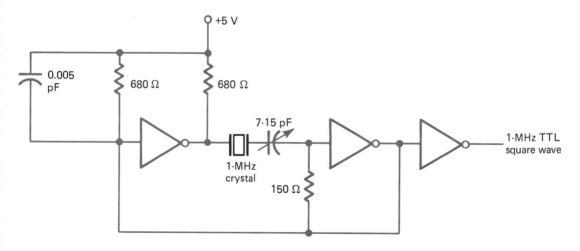

Fig. 9-4 A simple crystal oscillator. This crystal oscillator uses TTL integrated circuits. A crystal oscillator can also be built using bipolar or field-effect transistors.

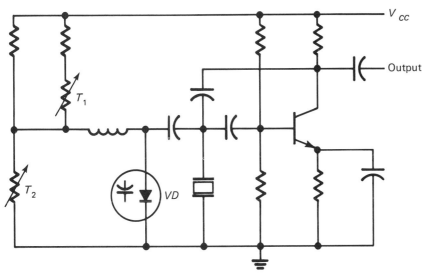

Fig. 9-5 The simplified schematic of a TCXO. The thermistor/varicap network keeps this oscillator on frequency with changes in temperature.

Ovened
oscillator

the temperature of the thermistor, the frequency of the oscillator is determined by temperature. As you can see, you must know the temperature characteristics of the crystal, the varicap, and the thermistors. Then a circuit can be built which returns the oscillator to the correct frequency when there are changes in temperature. TCXOs are the most common time-base oscillators used on good low-cost electronic counters.

There are limitations on how well the TXCO will stay on frequency with changes in temperature. If extreme stability is needed, the temperature effects on a crystal oscillator are removed by placing the crystal, and sometimes the whole oscillator circuit, in a temperature-controlled oven. The ovened oscillator is by far the most expensive, but definitely the most stable. Ovened oscillators can be made to almost any frequency stability you need. However, they are quite expensive. Electronic counters which use ovened oscillators are usually only the very expensive ones. The most complicated ovened oscillators have an ovened oscillator inside an oven. This double oven will hold the temperature to within 0.01°C for a major outside-temperature change. Such a double-oven system is usually reserved for very-high-performance laboratory counters. Ovened oscillators are ten to one thousand times better than a TCXO. They can cost hundreds to thousands of dollars.

Self Test

14. A typical crystal oscillator has a temperature coefficient of −1 ppm (parts per million) per degree Celsius. Its frequency is exactly 10,000,000 Hz at 25°C. It has a frequency of ___?___ Hz at 20°C.
    A. 9,999,950
    B. 9,999,995
    C. 10,000,005
    D. 10,000,050

15. A TCXO reduces the temperature coefficient of the time-base oscillator in self test question 14 to 0.1 ppm/°C. The temperature coefficient is now positive. At +30°C the oscillator's frequency is
    A. 9,999,950
    B. 9,999,995
    C. 10,000,005
    D. 10,000,050

16. A TCXO uses a thermistor, varicap, and crystal all matched to the circuit by computer. The varicap and the thermistor correct changes in the crystal frequency caused by a change in
    A. Time
    B. Temperature
    C. Supply voltage
    D. Shock

17. A double-ovened oscillator is held to within 0.01°C for an outside-temperature change of 10°C. The basic oscillator has a −10 ppm/°C temperature coefficient. The original oscillator frequency was

Decade counter

1,000,000 Hz. After an outside 10°C temperature increase, the oscillator frequency is now

A. 999,900 Hz
B. 999,990 Hz
C. 999,999 Hz
D. 999,999.9 Hz
E. 1,000,000.1 Hz
F. 1,000,001 Hz
G. 1,000,001 Hz
H. 1,000,010 Hz

The Divider Chain

As we noted in the time-base discussion, the time-base oscillator provides you with a frequency/time standard. Normally, this is an electronic oscillator operating at a frequency of 1 MHz. Often, you may wish to make

comparisons using standard times as long as 1 or even 100 s.

To obtain accurate signals over long periods, a divider chain is used. The divider chain is simply a series of decade counters. For example, the oscillator in Fig. 9-6 operates at a frequency of 1 MHz. Thus it has a period of 1 μs. The first decade counter divides this 1-MHz signal by 10. This gives a signal with a frequency of 100 kHz. This is a period of 10 μs. The second decade divider divides the 100-kHz signal by 10 resulting in a 10-kHz signal. This has a period of 100 μs. The third decade counter divides the 10-kHz signal by 10 producing a 1-kHz signal with a period of 1 ms.

Each additional stage of decade counting will divide the frequency by 10. As we can see

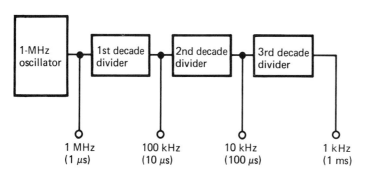

Fig. 9-6 A three-decade divider chain. Working with a 1-MHz oscillator, standard times of 1 μs, 10 μs, 100 μs, and 1 m S are generated.

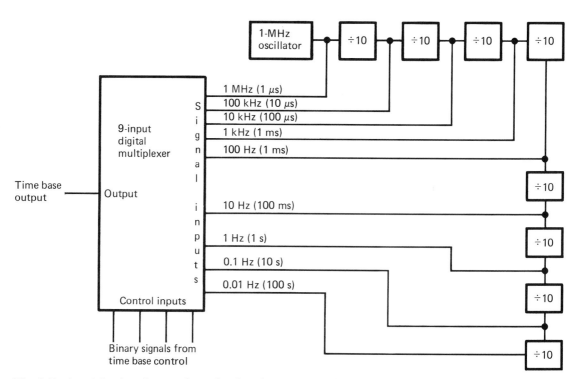

Fig. 9-7 An eight-decade time-base divider chain with multiplexer. Binary signals at the multiplexer's four-input control lines select the desired divider output signal.

from the above example, this increases the period by 10.

If the time-base oscillator operates at a frequency of 1 MHz, six decades of division give a 1-Hz (1-s) signal. Eight decades of division produce a 0.01-Hz (10-mHz) signal. This is a signal with a period of 100 s.

The simplified schematic in Fig. 9-7 shows an eight-decade divider chain. In addition to the decade dividers, a nine-input digital multiplexer is also shown. With the digital multiplexer you may select the division ratio you want from the divider chain. A digital multiplexer is used to give you remote-control operation. As you can see from the schematic, you do not actually switch the output of the divider chain; you only switch multiplexer control lines.

The divider chain does not change the time-base error. For example, a time-base oscillator with a frequency of 1,000,001 Hz is a 1,000,000-Hz (1-MHz) oscillator whose frequency is 1 Hz high. This is a frequency error of 0.0001%. This is often called an error of 1 ppm. The first decade divides the 1-MHz signal by 10. This gives a frequency of 100,000.1 Hz. The error is still 0.001% (1 ppm). When we examine the output after the sixth decade, we find that the frequency is 1.000001 Hz. This is still an error of 1 ppm. This almost 1-Hz signal has a period of 0.999999 μs. When the time-base frequency is higher than it is supposed to be, the period is shorter than it is supposed to be. Obviously, when the frequency is lower than it is supposed to be, the period will be longer than it is supposed to be.

Self Test

18. You are using a 10-MHz time-base oscillator. You select seven decades of division. The frequency at the output of the divider chain is
   A. 0.01 Hz
   B. 0.1 Hz
   C. 1 Hz
   D. 10 Hz
   E. 100 Hz

19. Five stages of decade dividing are used to reduce the frequency of a 1-MHz oscillator. The period of the waveform at the output of the divider chain is
   A. 100 s   D. 0.1 s
   B. 10 s   E. 0.01 s
   C. 1 s

20. A 10-MHz time-base oscillator is misadjusted to be 10,000,010 Hz. This means the 1-Hz output of the divider chain is actually
   A. 1.000010 Hz
   B. 1.000001 Hz
   C. 0.999999 Hz
   D. 0.999990 Hz

21. The period of the signal in self test question 20 is
   A. 1.000010 s
   B. 1.000001 s
   C. 0.999999 s
   D. 0.999990 s

The Gate and Control Circuits

When using the counter, you will either count pulses for an exact period of time or you will count exact time-base pulses for an unknown period of time. In either case you need a circuit which turns signals on and off as they flow to the decade counting and display unit.

The gate and control circuits perform this function. The gate and control circuits have two inputs: the signal input and the control input. Pulses at the signal input are passed to the decade counting and display unit when the correct pulses are connected to the control input. When the control input pulses let signal input pulses pass to the decade counting and display unit, the gate is said to be "open." When control input pulses stop signal input pulses from passing to the decade counting and display unit, the gate is said to be "closed."

When the gate closes, additional "housekeeping" pulses are generated. These housekeeping pulses are generated by the control circuits. The first two pulses generated are the memory and the reset pulses. These are normally generated one after another when the gate closes.

First, the memory pulse is generated. This pulse commands the transfer of information in the decade counters to the latches. Second, once the information is stored in the latches, a second pulse from the control circuit resets the decade counters. That is, the count in the decade counters is set to zero. The decade counting and display unit is now ready for a new measurement to take place.

The normal use of the electronic counter is to display numbers meaningful to people. To do this, the display must not change too often.

Digital multiplexer

Time-base error

Open gate

Closed gate

Time out pulses

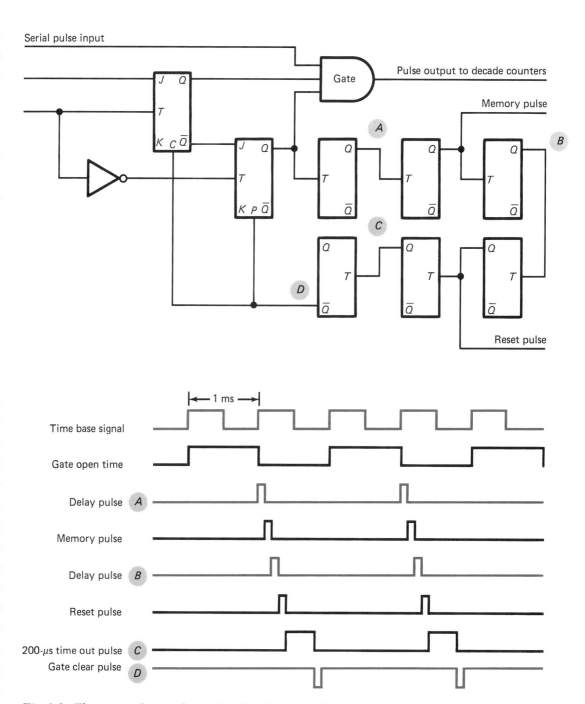

Fig. 9-8 The gate and control circuit. Signals to the flip-flops open and close the gate. A series of monostable multivibrators generate the memory, reset, time out, and delay pulses.

The gate and control circuits also make sure that measurements are not made faster than about five times a second. If the measurements are made faster than five times per second, the human eye cannot follow the changing digits. The gate and control circuits add *time out* pulses. These time out pulses make the instrument only take five readings per second. Sometimes, you will want to slow mea-

surements to less than five per second. The more expensive electronic counters allow you to adjust the time between measurements.

The schematic in Fig. 9-8 shows a simple gate and control circuit which uses integrated-circuit monostable multivibrators. This figure also includes the waveform chart. The waveform chart shows how each monostable is fired one after another. The first monostable is triggered by the gate closing. We can see that once the gate closes, the memory, reset, and time out pulses are generated one after another. Each pulse is separated from the signal which caused it by a delay pulse. Once the time out pulse is over, the gate can be opened again.

Self Test

22. Some electronic counters have a switch which lets you see the decade counters count. This switch controls the _____?_____ pulse.
    A. Memory
    B. Reset
    C. Time out
    D. Input

23. All decade counters in an electronic counter begin a measurement by being set to zero counts. This is done by the _____?_____ pulse.
    A. Memory
    B. Reset
    C. Time out
    D. Input

24. The gate passes _____?_____ pulses to the decade counting and display unit when it is open.
    A. Memory
    B. Reset
    C. Time out
    D. Input

25. If an electronic counter's display changes more than five times per second, it is hard for the human eye to follow the changes. The _____?_____ pulse makes sure the counter does not take more than five readings per second.
    A. Memory
    B. Reset
    C. Time out
    D. Input

The Input Amplifier and Shaping Circuits

The input amplifier and shaping circuits appear to be some of the simplest circuits in an electronic counter. These circuits connect the low-level real-world signals to the digital integrated circuits. The shaping circuits drive the gate and control circuits.

You may divide these analog circuits in two basic functions. First, the signal must be am-

plified before it can drive the shaping circuits. The shaping circuits do not amplify the signal. They change the rounded edges of waveforms, such as the sine wave, into signals with sharp leading and trailing edges, like the square wave or pulse.

Figure 9-9 shows a simplified schematic diagram of an input amplifier and shaping circuit combination. The input signal first goes to an input attenuator. This attenuator reduces the input signal to a level which will not cause the amplifier to distort. Of course, no attenuation is used for the lowest input signal levels. The input signal is then amplified by the input amplifier.

When the input signal is amplified and shaped, some other changes may be made to the signal. For example, many counters have ac/dc coupling. With ac coupling the signal is connected to the attenuator through a capacitor. It is connected directly with dc coupling. This is done by switch S_1 in Fig. 9-9. If the input is connected through the capacitor, only ac signals of approximately 10 Hz or greater reach the input amplifier and shaping circuits. However, with direct coupling, dc signals control the output of the shaping circuits.

A second feature which is often used is a level control. This allows you to select the point on the input waveform where the shaping circuits trigger. It is shown in Fig. 9-10. By adjusting the level control, you may set the *trigger point*. This can be anywhere from the negative peak of the incoming signal to the positive peak of the incoming signal. The level control is very useful when the input signal has a *glitch* or an unwanted pulse. Such a glitch is diagrammed in Fig. 9-11.

This figure shows two settings of the level control. The first setting makes the shaping circuits trigger when the signal passes through 0 V. This makes an extra pulse at the output of the shaping circuits. The extra pulse is the point where the glitch makes the input waveform pass back through 0 V. The second setting of the level control does not allow the glitch to pass the input waveform back through the trigger point; so the glitch is gone.

Counters which have a trigger level control also have a trigger level switch. This switch (S_3 in Fig. 9-9) connects the inverted or noninverted signal from the amplifier output to the input of the shaping circuit. Figure 9-12(*a*) and (*b*) shows what the trigger level switch does to the output waveform. The level con-

Error from noise

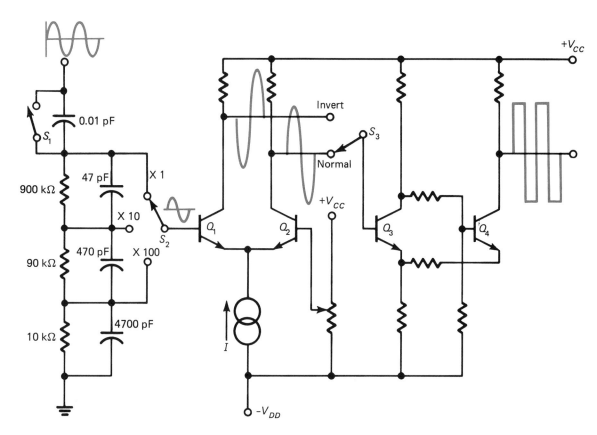

Fig. 9-9 The input amplifier and shaping circuit. The input amplifier is made up of the differential transistor pair Q_1 and Q_2. It has gain and gives both an inverted and a noninverted output. The shaping circuit is a Schmitt trigger made up of Q_3 and Q_4.

trol and the level switch are used together. The level control lets you trigger anywhere over a 180° portion of the input waveform. The level switch allows you to select which 180° section of the input waveform you wish to trigger on. This is shown in Fig. 9-12(c).

Although the input circuits are extremely useful, they do introduce measurement errors. Most of the error comes when noise on the input signal reaches the shaping circuits. Figure 9-13(a) shows the output of the shaping circuits with an ideal input signal. In this case, an ideal input signal has no noise. In Fig. 9-13(b), you can see the output of the

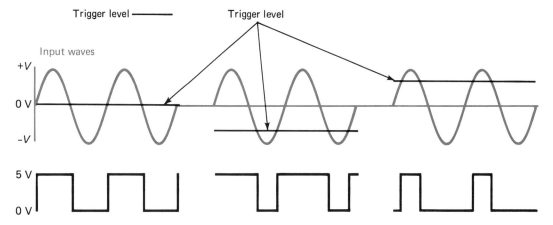

Fig. 9-10 Using the level control. The level control adjusts the differential amplifier dc offset. This control lets you select the best trigger point on the input waveform.

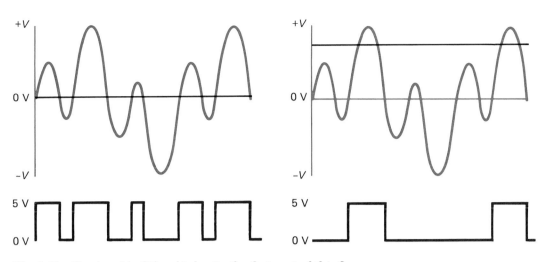

Hysteresis

Fig. 9-11 Getting rid of the glitch. In the first part of this figure the level control is set to trigger at zero crossing. This gives a double output pulse. In the second part, the level control is adjusted to give a single trigger point for each cycle of the input waveform.

shaping circuits when the input signal has a great deal of noise on it. Note the first pulse and the second pulse of the output waveform are different widths because of the noise. The design objective of any input amplifier/shaping circuit combination is to not add noise to the incoming signal. It is also to stop the noise from changing the output signal if at all possible. If the noise is too great, it looks like the glitch we discussed earlier.

To avoid this problem, hysteresis is introduced on the shaping circuits. Hysteresis means that the input signal voltage needed to make a positive-going output from the shaping circuit is greater than the voltage needed to cause a negative-going output. This is shown in Fig. 9-14.

With hysteresis in the circuit, normal amounts of noise will not cause additional output pulses. Hysteresis, however, reduces the

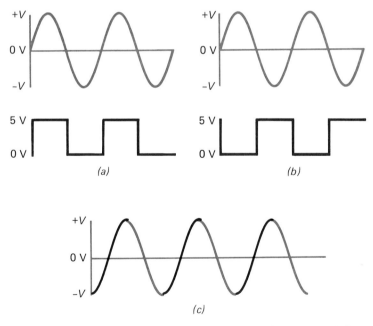

Fig. 9-12 (a) The trigger level switch is set to the positive (+) level. (b) The trigger level switch is set to the negative (−) level. (c) The black portion of the waveform shows the + level triggering range. The colored part of the waveform shows the − level triggering range.

Instrument sensitivity

Schmitt trigger

Voltage comparator

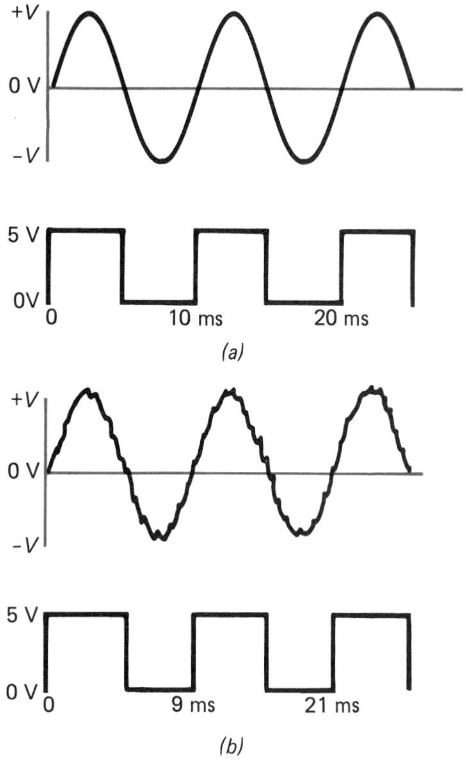

Fig. 9-13 Input noise. (*a*) The response of a shaping circuit to a "clean" signal. (*b*) The response of a shaping circuit to a noisy signal.

basic sensitivity of the instrument. This is because signals with an amplitude less than the hysteresis will not trigger the shaping circuits. This can be seen in Fig. 9-14.

In this figure, a signal of ±1 V triggers the

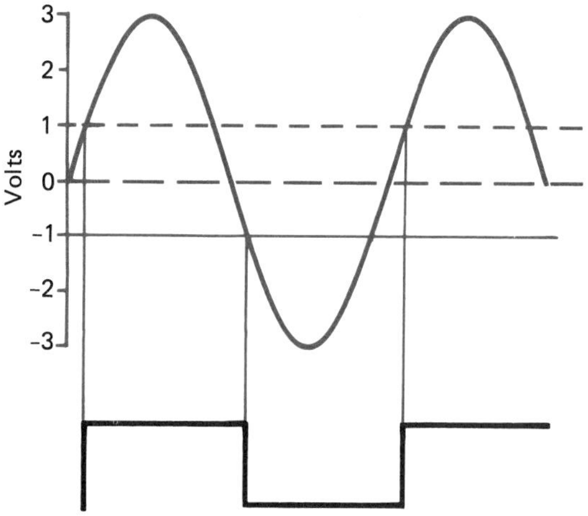

Fig. 9-14 Hysteresis. Hysteresis is added to the shaping circuits to keep them from triggering on noise. The 0 to 1 change comes as the input signal swings positively through +1 V. The 1 to 0 change does not happen until the signal swings negatively through −1 V. Signals which do not reach from +1 V to −1 V will not trigger the shaping circuit at all.

shaping circuit. But a signal less than ±1 V will not trigger it. If this shaping circuit's amplifier has a gain of 25, the sensitivity of the electronic counter is

$$\left(\frac{\pm 1\ \text{V}}{25}\right) \times (\pm 40\ \text{mV})$$

That is, signals of less than ±40 mV at the input will not trigger the shaping circuits. Therefore, they will not operate this electronic counter. You can see that a compromise between noise-rejection capability and sensitivity must be reached.

The circuits used for wave shaping are either the Schmitt trigger or the voltage comparator. Both circuits do the same job. Most new counters use integrated-circuit voltage comparators for the shaping circuit. Older counters use Schmitt triggers. Frequently both the input amplifier and the voltage comparator are built using ECL (emitter-coupled logic). This logic has excellent high-frequency response capabilities and can be operated in the linear as well as the digital mode.

Self Test

26. One particular shaping circuit has a ±0.5-V hysteresis. An amplifier with a gain of 50 precedes the shaping circuit. The sensitivity of this counter is ___?___ mV peak-to-peak.
    A. ±500
    B. ±100
    C. ±50
    D. ±10

27. An electronic counter is set to trigger on the positive-going slope. The input/output waveforms are shown in Fig. 9-15. What do the input/output waveforms look like if the level switch is set to trigger on the negative-going slope?

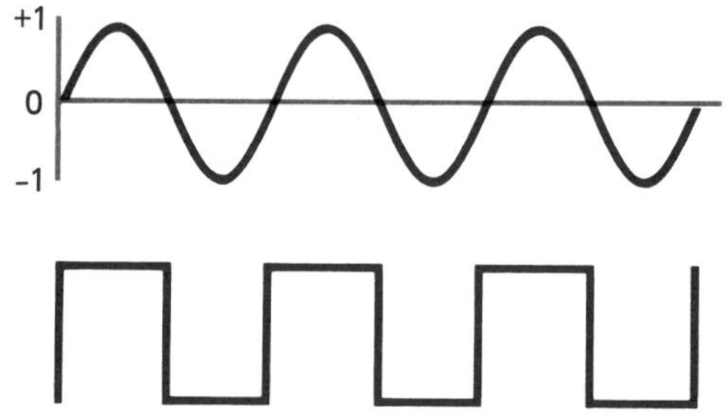

Fig. 9-15 Self test question 27.

28. The trigger level control in self test question 27 is set to trigger at
    A. −1 V
    B. −0.5 V
    C. 0 V
    D. +0.5 V
    E. +1 V

29. The waveform in Fig. 9-16 has a glitch which causes double counting. Show by a horizontal dotted line where the trigger level control must be set to avoid this glitch and yet still trigger the shaping circuits.

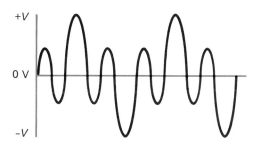

Fig. 9-16   Self test question 29.

30. The input/output waveforms in Fig. 9-17 show a dc-coupled signal triggered at +1 V. Show the output waveform if the signal is ac-coupled.

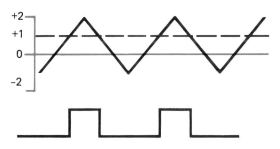

Fig. 9-17   Self test question 30.

9-4 CONVERTING THE BUILDING BLOCKS INTO INSTRUMENTS

We have studied the operation of the digital and analog building blocks used to construct an electronic counter. Some instruments do not use all the building blocks. Others use more than one of some building blocks. In either case, these blocks work just as we explained. How we use them does not change how they work.

The Events Counter

The simplest electronic counter is the events counter. The events counter counts how many pulses happen over a selected period of time. The block diagram of an events counter is shown in Fig. 9-18.

The events counter does not use the time base or the divider chain.

The source of pulses you wish to count is connected to the counter's input. After these pulses are amplified and shaped, they pass through the gate. The gate is controlled from a start/stop switch on the instrument's front panel. From the gate the pulses go directly to the decade counting and display unit. A reset button sets all decade counters to zero.

Counting begins when you open the gate by setting the switch to Start. When you flip the switch to Stop, closing the gate, the counting ends. The display is then updated by a memory pulse. The display shows the number of pulses counted over the time you selected.

As you can see, the events counter is a very simple instrument. A second amplifier/shaping circuit may be used to open and close the gate instead of the switch. This adds to the usefulness of an events counter, since it is not necessary to manually press a switch to control the opening and closing of the gate.

Start/stop switch

Self Test

31. You open the gate to your events counter. Sixty pulses come in during the first ½ s, 30 in the next ½ s, and none for the next second. You then shut the gate. The display reads
    A. 2
    B. 0.5
    C. 45
    D. 90

32. One advantage the electronic events counter has over the mechanical counter is its speed. The electronic events counter can count millions of pulses per second if needed. Another advantage is that
    A. It has built-in signal amplification and shaping
    B. Only electronic instruments can be read in the dark
    C. Electronic counters have at least eight digits
    D. It has an electronic reset

105

EPUT counter

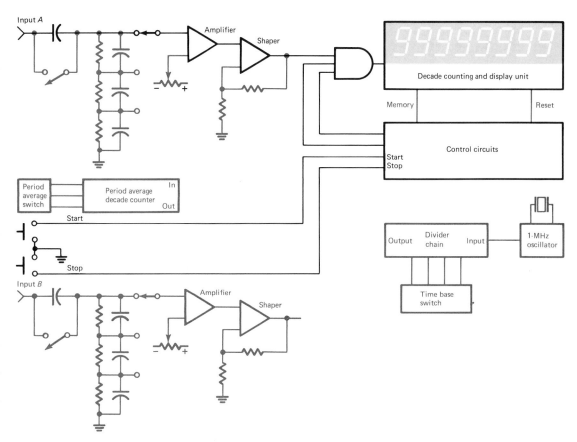

Fig. 9-18 A simplified block diagram of the events counter.
Pulses from the input go to the decade counting and display unit
through the gate when the start button is pressed.

33. An events counter has nine digits. This
means it will not overrange until at least
_____?_____ counts are totalized.
A. 1 million
B. 10 million
C. 100 million
D. 1 billion

34. You are trying to find the rpm of a slowly
turning shaft. You get two pulses every
time the shaft goes around once. You
turn on your events counter for exactly
1 h. The display reads 900 at the end of
the measurement time. The shaft is
turning at
A. 900 rpm
B. 450 rpm
C. 7.5 rpm
D. 15 rpm

35. An events counter could not be used to
directly measure
A. Gallons per hour
B. A production-line daily output
C. Voltage
D. The number of cars which pass one
stoplight in one day

The Frequency Meter

The introduction of the electronic counter
produced a new way of thinking about frequency. The common definition of frequency tells us it is the number of cycles in 1 s.
In digital circuits, we can think of this as the
number of pulses in 1 s. The first electronic
counters were simply events counters. Of
course, each event is one pulse, one cycle of
an unknown signal.

Early users of the electronic counter soon
realized that events counted over a standard
time interval (such as 1 s) gave a display
directly in hertz. For example, if 1000 events
(pulses or cycles) are counted while the gate is
open for 1 s, the frequency of the signal is
1000 Hz. This type of thinking leads to the
name events per unit time (EPUT).

The original EPUT counter was the first
digital frequency counter. When you think of
using the events counter to measure frequency, you think of opening the gate for exactly 1 s. This is because we are all familiar

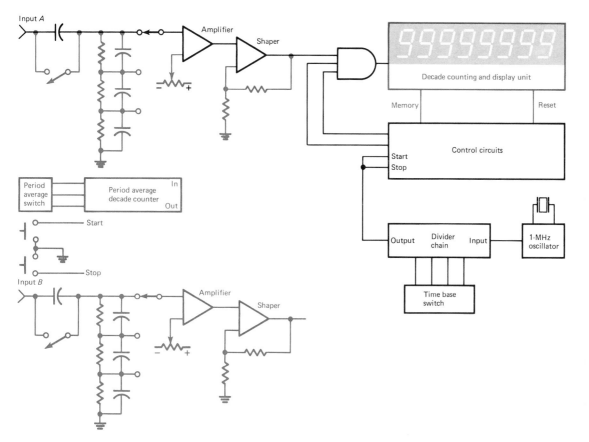

Time-base

1-s gate interval

Fig. 9-19 The frequency counter. Pulses from the input go to the decade counting and display unit when the gate is open. The gate is open for the exact time you select from the time base.

with the normal definitions for hertz, kilohertz, and megahertz. We talk about 1 cycle per second, 1000 cycles per second, and 1 million cycles per second.

In Fig. 9-19 the blocks of the electronic counter are put together as a digital frequency meter. You can see this is very much like the events counter. The only difference is how the gate is opened and closed. With the events counter, you press a switch to open and close the gate. With the digital frequency meter, you must open the gate for exactly 1 second. If the gate opens for less than 1 second, too few counts go to the decade counting and display unit. This will cause a low frequency reading. If the gate is opened too long, too many counts go to the decade counting and display unit. The frequency reading will be too high.

The time-base oscillator and the divider chain are used to be sure the gate is opened for exactly 1 s. In this case, the exact 1-Hz (1-s) signal is selected. This 1-Hz signal is connected to the gate and control circuits. The

output of the divider chain now has one leading edge every second. The first leading edge opens the gate. The second leading edge closes the gate. After the second leading edge there is a 1-s pause until the third leading edge comes from the divider chain. The third leading edge opens the gate again. The fourth leading edge closes the gate again. The waveforms in Fig. 9-20 show how these signals work.

In this example, the gate is opened for exactly 1 s. When a 1-s gate interval is used, one measurement is made every 2 s. Sometimes you will want to make measurements faster than one every 2 s. Using the 10-Hz tap on the divider chain we can open the gate for exactly 0.1 s. We then wait 200 ms for the gate control circuits to time out. The gate can then be opened for another 0.1 s. Doing this allows us to make readings as fast as the control circuits in the gate permit.

Frequency is defined as cycles per second. It is not defined as cycles per tenth of a second. Fortunately, a tenth of a second is related to

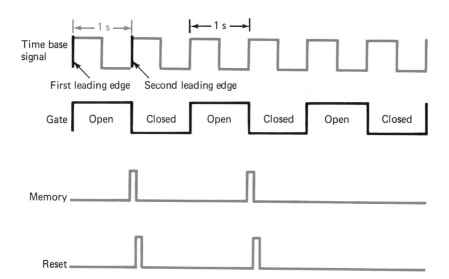

0.1-s gate
interval

Measurement
resolution

Fig. 9-20 The control signals in a frequency meter. The gate opening is controlled by the exact signals from the time-base oscillator. Memory, reset, and time out pulses come just after the gate closes.

1 s by a factor of 10. We can correct for a 0.1-s gating interval by simply moving the decimal point on the display.

For example, suppose your frequency meter is connected to a 3.5-MHz oscillator and it has a gating interval of 1-s. There will be 3,500,000 pulses counted in 1 s. If a 0.1-s gating interval is used, only 350,000 pulses are counted. In this case, we may show the frequency in kilohertz. Using a 1-s time base, we can display the frequency as

$$3500000 \text{ Hz}$$

When using the 0.1-s time base, we can display the frequency as

$$3500.00 \text{ kHz}$$

The first measurement has a resolution of 1 Hz. The second measurement has a resolution of only 10 Hz. Otherwise, the displayed readings are exactly the same. In other words, we have traded an increase in measurement speed for a loss in resolution.

Just as we can make frequency measurements using a 0.1-s gate time, we can use other gate times. They just need to be decade multiples (or submultiples) of 1 s. All we have

to do then is shift the decimal point to the right or to the left a few places to make the display read correctly. Commonly used gating intervals for frequency meters are 1 ms, 10 ms, 0.1 s, 1 s, 10 s, and 100 s.

The chart in Fig. 9-21 shows these gating times and the resolution each one gives. As you can see by looking at this chart, the longer the gate time the greater the resolution. Of course, you always must ask yourself, "Can I use this increased frequency resolution?"

For example, suppose you know the time-base oscillator on a counter is accurate only to 1 ppm. In this case, there is no point in mea-

Gate time	Resolution	Measurement Time*
1 ms	1000 Hz	5 per second
10 ms	100 Hz	5 per second
100 ms	10 Hz	3 per second
1 s	1 Hz	1 every 2 seconds
10 s	1/10 Hz	1 every 20 seconds
100 s	1/100 Hz	1 every 200 seconds

*200 ms holdoff presumed

Fig. 9-21 Gating intervals and resolution. Faster frequency measurements mean lower resolution. Some counters offer only two gating intervals, 1 ms and 1 s.

suring a 10-MHz signal for 10 s. This measurement would give you 0.1-Hz resolution. This is 0.01 ppm. The frequency resolution of the measurement is much greater than the frequency accuracy of the instrument.

One of the most common questions about a frequency meter is, "What is the highest frequency which it will count?" The frequency meter is counting events per unit of time. These events (or pulses) must pass through the input amplifier, the shaping circuits, the gate, and the first decade counter. All these sections must work at the highest frequency you wish to count. As you can see, this controls how the input amplifier and shaping circuits are designed. It also controls what logic families are used to construct the gate and decade counter.

If normal transistor-transistor logic (TTL) is used, upper frequencies in the 30-MHz range may be expected. If Schottky TTL is used, upper frequencies in the 80-MHz range may be expected. For frequencies greater than 80 MHz, emitter-coupled logic (ECL) is used. Of course, this same rule limits the rate at which we can count events. The rate at which events are counted can be no faster than the maximum frequency capability of these circuits.

Self Test

36. You select a 1-ms gate interval to measure a 52-MHz signal. The counter will display this as
   A. 52,000,000 Hz
   B. 52,000.00 kHz
   C. 52,000.0 kHz
   D. 52,000 kHz

37. If the gating interval is longer than indicated, the counter will display
   A. The correct frequency
   B. A count which is too high
   C. A count which is too low
   D. The exact time-selected time base

38. The first decade counter of a digital frequency meter is built using 160-MHz emitter-coupled logic. The rest of the instrument is built with 30-MHz TTL. The instrument's maximum frequency capability is
   A. 160 MHz
   B. 30 MHz
   C. 80 MHz
   D. 15 MHz

39. You are trying to set the master oscillator in an electronic organ to 440 Hz ±0.01 Hz. To get this resolution, you use a ____?____ gating interval.
   A. 100-ms
   B. 1-s
   C. 10-s
   D. 100-s

40. Complete the table below to show how a seven-digit electronic counter would display a 144.6-MHz signal with the different gate intervals. Also indicate the resolution you would expect to get.

Gate Interval	Resolution	Reading	Units
1 μs			MHz
10 μs			MHz
100 μs			MHz
1 ms			kHz
10 ms			kHz
100 ms			kHz
1 s			Hz
10 s			Hz

The Period Meter

Frequency and time are related by the equation

$$T = \frac{1}{f}$$

Often when we use electronic instrumentation we switch between frequency and time measurements.

For example, a signal with a period of 1 ms has a frequency of 1 kHz. A signal with a period of 10 ms has a frequency of 100 Hz. In some measurements frequency gives the most resolution. In other measurements time gives the most resolution. Fortunately, most electronic counters give either the frequency or the period of a signal. This allows you to select the measurement which is most useful.

Frequency is measured by counting the number of unknown pulses which happen in the standard time. Period is the opposite of this. It is the number of standard time pulses which happen in one cycle of the unknown signal.

The block diagram of Fig. 9-22 shows how the electronic counter's blocks are hooked up

Accuracy versus resolution

Maximum frequency

Transistor-transistor logic (TTL)

Schottky TTL

ECL

Resolution

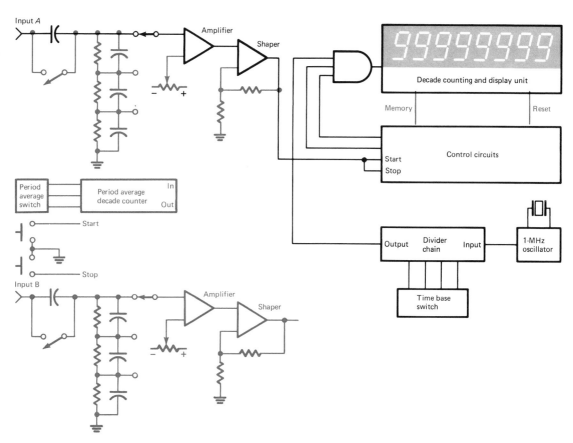

Fig. 9-22 The period meter. Pulses at a known rate go to the decade counting and display unit when the gate is open. The gate is open during the period of one cycle of the input signal.

to make a period meter. There is only a small difference between the frequency meter and the period meter. The difference is how the input and time-base signals are connected to the gate and control circuits. In the period meter, they are opposite to the connections used for frequency measurement. The control signals are shown in Fig. 9-23.

As you can see in the period-meter block diagram (Fig. 9-22), the time base is connected to the gate input. The input signal controls the time the gate is open. Let's now look at some examples of how the period meter works.

Assume, for example, we have a period meter connected to a 5-Hz signal source. The period of this signal is

$$T = \frac{1}{f} = \frac{1}{5 \text{ Hz}} = 0.2 \text{ s} = 200 \text{ ms}$$

This is to say, the period of the 5-Hz signal is 0.2 s. This is the same as 200 ms or 200,000 μs. The leading edge of the first 5-Hz

cycle is used to open the gate. The leading edge of the next 5-Hz cycle is used to close the gate. The gate is then open for a time exactly equal to one period of the 5-Hz waveform. The exact time-base pulses go to the decade counting and display unit during this time.

Suppose we select the 1-MHz (1-μs) time-base signal. While the gate is open, the counting and display unit gets one pulse every microsecond. The gate is open for 200,000 μs; so 200,000 pulses go into the decade counting and display unit. When the gate closes, the display reads 200,000. This shows the period of a 5-Hz waveform is 200,000 μs.

Like the frequency meter, a different output from the divider chain may be used. For example, assume the multiplexer is set to select the 1-ms (1-kHz) output from the divider chain. If the gate is open for 200,000 μs (200 ms), the counting and display unit will get 200 1-ms pulses. This will be displayed as 200 when the gate closes. This is read as a period of 200 ms.

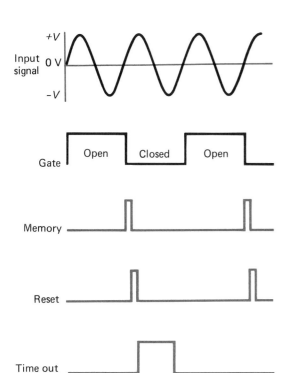

Fig. 9-23 The control signals in the period meter. The gate opening is determined by the signal applied to the input circuit. Memory, reset, and time out pulses come just after the gate closes.

If you use the 1-MHz output from the divider chain, the resolution is 1 μs. If, for example, you change the output of the divider chain to the 100-μs position (10 kHz), the resolution will be 100 μs (0.1 ms).

Period measurements are very useful when low frequencies are being measured. For example, assume you wish to know the ac power-line frequency accurately. You can make a frequency measurement with a frequency counter. You can use a 100-s gate interval to measure the line frequency with a resolution of 0.01 Hz. Another way to say this is that the resolution is 0.016%. On the other hand, if you measure the period of the line signal to the nearest microsecond, this gives a resolution of 0.006%. This is an improved resolution by a factor of almost 3.

You can get greater improvements if the frequency is lower. For example, a 1-Hz signal can be measured to the nearest 0.01 Hz (1%) using a 100-s time base. However, using a 1-μs period measurement, it may be measured to the nearest 1 ppm (0.001%). This is a significant improvement. In fact, this improvement is so great that other errors on the input signals now prevent you from making

these period measurements to this accuracy. These errors are discussed later.

Self Test

41. You measure the period of the ac power line with a resolution of 1 ms. Your counter reads
A. 60
B. 8.334
C. 16.667
D. 30
E. 8
F. 16

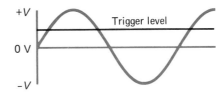

Fig. 9-24 Self test question 42.

42. By adjusting the trigger level control from zero crossing to the point shown in the waveform of Fig. 9-24, the new period measurement
A. Remains the same as the first measurement
B. Drops to one-half of the first measurement
C. Drops to 0.77 of the original measurement
D. Increases to two times the original measurement
E. Increases to 1.414 times the original measurement

43. Complete the following table for a 0.1-μs resolution period measurement of a 0.1-Hz signal on a seven-digit period meter.

Time Interval	Resolution	Reading	Units
1 μs			μs
10 μs			ms
100 μs			ms
1 ms			ms
10 ms			s
100 ms			s
1 s			s
10 s			s

111

44. You are trying to get a 2-Hz oscillator set to within ±0.1%. How long does the frequency measurement take? How long does a period measurement take?

The Time-Interval Meter

Period measurements are used to measure the time needed for a signal to complete one full cycle. The period measurement gives the time between two identical points on the waveform. There is one other important timing measurement. It is called a time-interval measurement. The time-interval measurement is the time between a start signal at one input of the electronic counter and a stop signal at a second input of the electronic counter. The input signals can be taken from one signal source or from two different signal sources.

For example, Fig. 9-25 shows a setup used to measure the speed of a bullet. The start signal is connected to a switch which closes when the bullet is fired from the gun. The stop signal comes from a switch 100 ft away. This switch closes when the bullet passes it. Once the time needed for the bullet to go 100 ft is known, the speed of the bullet can be simply computed by using the formula

$$\text{Velocity} = \frac{\text{distance}}{\text{time}}$$

Figure 9-26 is a time-interval meter block diagram. As you can see, the time-interval meter has almost the same block diagram as the period meter. There is one difference.

The input signals for the gate and control circuits come from two amplifiers and shaping circuits. In the period meter they came from just one amplifier and shaping circuit. The gate is opened by a signal on the instrument's A input. It is closed by a signal that you apply to the instrument's B input.

As we noted before, these two signals may be from completely independent sources or may be connected to the same source as shown in Fig. 9-27. In this case, the A signal is set to trigger on the leading edge of the pulse. The B signal is set to trigger on the trailing edge of the same pulse. This allows you to make a pulse-width measurement. In this particular case, the width of a positive pulse is measured. However, the width of a negative pulse may be measured if the A input is set to trigger on the trailing edge of the pulse and the B input is set to trigger on the leading edge of the pulse.

If the input amplifier and shaping circuits have level controls, you may also select the point on the leading or trailing edge at which the measurement will start and stop. If you have this control, you may make a measurement such as the one shown in Fig. 9-28. Here a time-interval measurement is made. This time-interval measurement measures the rise time of the pulse. The rise time is the time required for the pulse to go from 10 to 90% of its amplitude.

In this example, both the A (Start) and the B (Stop) inputs of the time-interval meter are connected to the pulse generator's output. Both inputs are set to trigger on the leading edge of the pulse. Input A is set to trigger at

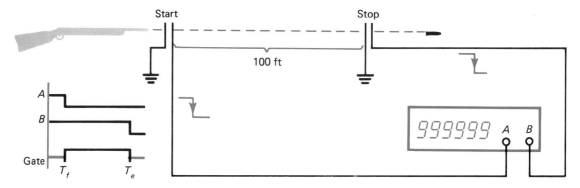

Fig. 9-25 Measuring speed. When the bullet goes through the start detector, a pulse is sent to the A input of the time-interval meter. This opens the gate. When the bullet goes through the stop detector 100 ft away, a pulse is sent to input B of the time-interval meter. This closes the gate, ending the measurement.

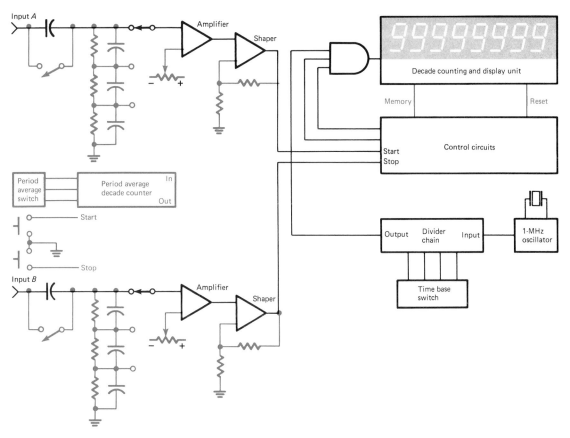

Fig. 9-26 The time-interval meter. Pulses from the time base go to the decade counting and display unit from the time the gate is opened by a pulse on the start input until the gate is closed by a pulse on the stop input.

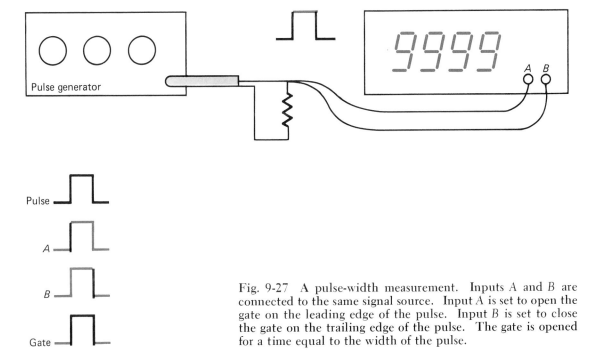

Fig. 9-27 A pulse-width measurement. Inputs A and B are connected to the same signal source. Input A is set to open the gate on the leading edge of the pulse. Input B is set to close the gate on the trailing edge of the pulse. The gate is opened for a time equal to the width of the pulse.

113

Resolution

Multiple-period
meter

Error from noise

Signal-to-noise
ratio

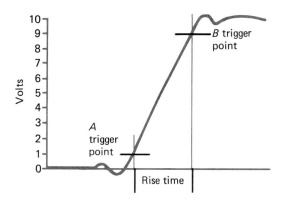

Fig. 9-28 A rise-time measurement. Using the trigger level controls, the start/stop points are set at +1 V and +9 V on the leading edge of a 10-V pulse. The time from +1 to +9 V is the pulse's rise time.

the 10% point (1 V). Input *B* is set to trigger at the 90% point (9 V). An oscilloscope or some other measurement device is necessary to set the *A* and *B* level controls to the exact voltage needed. The time-interval meter shows the time for the pulse to rise from 10 to 90%. That is to say, the measurement is a rise-time measurement.

As you can see from the above discussion, the time-interval meter has the same resolution as the period meter. The only real difference between the two is how the gate gets its open/close signals. The time-interval meter provides much more flexibility.

Self Test

45. You measure a positive pulse width of 10 μs using a time-interval meter. When you flip the electronic counter to the period function, you read 100 μs. Therefore, the negative pulse width must be
   A. 10 μs
   B. 90 μs
   C. 100 μs
   D. 900 μs

46. You wish to measure the time that the output of a dual-slope converter's integrator is above 0 V. To do this you use a
   A. Frequency measurement
   B. Period measurement
   C. Time-interval measurement
   D. Events measurement

47. Radar measures distance by determining how long it takes a pulse to travel to an object and back again. This can be measured digitally by using a

A. Frequency meter
B. Period meter
C. Time-interval meter
D. Events counter

48. Your electronic counter has a 1-MHz time-base oscillator. The minimum pulse width you can measure with 10% or better accuracy is a _____?_____ pulse.
   A. 1-μs
   B. 10-μs
   C. 100-μs
   D. 1-ms
   E. 10-ms
   F. 100-ms
   G. 1-s

The Period-Average Meter

The period-average meter is a special period function that may be found on some counters. The period-average meter is sometimes called the multiple-period meter. It is a function designed to get rid of the most common errors you find when making period measurements. The period-average meter displays the average time for a number of periods. This average is used to remove the error.

The error in a period measurement comes from noise on the input signal. For example, looking at Fig. 9-29, you can see the width of the gate-interval signal changes because of the noise on the input signal. Although it is possible to reduce the noise to a great degree, it is impossible to get rid of all of it.

For example, let us look at a 1-V signal which has 0.01 V of noise. You can see the ratio of the signal to the noise is 100 to 1. This is called a *signal-to-noise ratio* of 40 dB. A 40-dB signal-to-noise ratio is quite common. A 60-dB signal-to-noise ratio (the signal is 1000 times greater than the noise) can be found, but it is not common.

It can be shown that a 40-dB signal-to-noise ratio on a sine wave can cause a period-measurement error (a gating pulse with jitter) of ±0.3%. Needless to say, this is a significant error when you think of the capabilities of, for example, a six- to nine-digit digital instrument.

Let us now look at the effects of period averaging. The error in a period measurement is due to noise in the input signal. When we are measuring a single period, this noise affects the entire measurement.

Suppose we redesign the period meter so the gate is open for 10 periods of the unknown

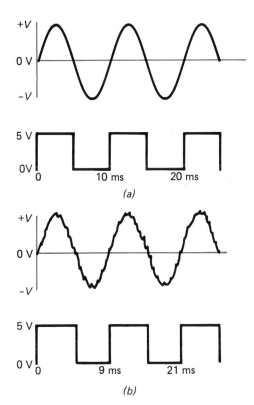

Fig. 9-29 Noise error. (a) A clean sine wave produces pulses of equal width. (b) The shaped pulses are not the same width because of the noise on the input signal.

signal you are measuring. When we measure 10 periods, only the start of the first period and the stop of the last period have the 0.3% error. All the other periods are measured with no error. This is because the input gate remained open for all those periods. The error in this measurement is reduced by a factor of 10.

Obviously, you can extend this technique to any number of periods you desire. Period-average measurements are often made over 1000 periods. Thus a normal error, which is ±0.3% on a 40-dB signal-to-noise ratio, is reduced to 0.0003%! This error is much more like the error we expect from a digital instrument.

What do we give up to gain this accuracy? To gain accuracy in a period measurement, we must take more time for the measurement.

For example, if we are measuring the period of a 500-Hz oscillator, the actual measurement takes only 0.002 s (2 ms). This is the period of the 500-Hz signal we are measuring. However, to improve the accuracy by a factor of 1000, we must average 1000 periods. Therefore, the measurement time is extended from an insignificant 2 ms to 2000 ms (2 ms per

period × 1000 periods). It now takes 2000 ms (2 s) to make the same measurement. But you get 1000 times more accuracy.

Figure 9-30 shows a simplified block diagram of a period-average meter. You can see this block diagram is almost exactly the same as the block diagram for a period meter. The only difference is the decade counters which are added to keep track of how many periods have been averaged. Obviously there are some additional simple circuits to keep track of the decimal point as you change the number of periods which are to be averaged.

Self Test

49. You want to set a 440-Hz oscillator to within 0.01 Hz. You know this is a period of 2.272727 ms. Which method gives you the quickest measurement in time: period average or frequency? Why? What is the required gate interval time for a frequency measurement? How many periods do you have to average?

50. The period-average function requires some extra decade counters to be added to the instrument. These are to
    A. Update the displays
    B. Count the time-base pulses
    C. Count the number of periods being measured
    D. Generate reset pulses

51. A signal with a 40-dB signal-to-noise ratio has an error in the period mode of
    A. 1.0%
    B. 0.3%
    C. 0.1%
    D. 0.03%

52. A counter has a time-base oscillator which has only ±15 ppm accuracy. It does 10,000 period averages. Which causes the most error, the time-base error or a 40-dB signal-to-noise ratio period-average measurement? Why?

53. Period averaging helps reduce error because the noise affects only
    A. The first cycle
    B. The last cycle
    C. Every cycle
    D. The first and last cycles

54. Period averaging is also known as
    A. Multiple period
    B. Frequency ratio
    C. Time interval
    D. Period

115

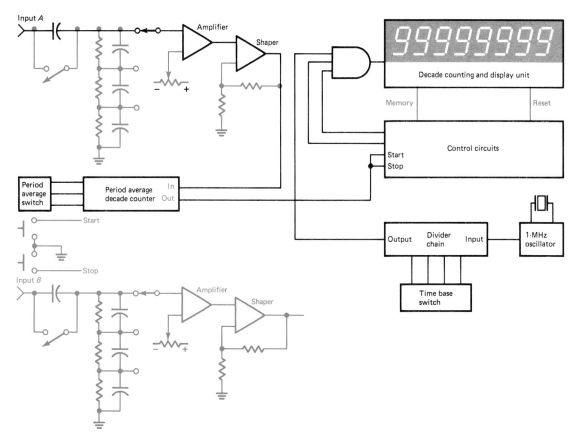

Fig. 9-30 The period-average meter. Pulses at a known rate go to the decade counting and display unit when the gate is open. The gate is open for a selected decade number of periods of the input signal.

Summary

1. The job of the electronic counter is measuring frequency and period, or time. Frequency is the number of complete cycles occurring in 1 s. Period is the time required to complete one whole cycle. Frequency f and period T are related by the formula

$$T = \frac{1}{f}$$

2. The heart of the electronic counter is the decade counting and display unit. This unit also has the overrange circuit.

3. The time-base oscillator gives the electronic counter a stable frequency or time reference. The quartz-crystal oscillator is used because it is stable and inexpensive.

4. Electronic counters need a number of high-accuracy reference frequencies. These are generated by dividing the time-base oscillator, by means of a divider chain.

5. A digital multiplexer is used to remotely select the desired divider-chain output. Each decade division reduces the frequency and period by 10.

6. The gate and control circuits determine when pulses are to go to the decade counting unit and display units. The control circuit generates memory, reset, and time out pulses when the gate closes.

7. The input amplifier provides the gain needed to bring the signal up to integrated-circuit voltage levels. The shaping circuits change the input waveforms into a pulse train. The input attenuator reduces the input signal to prevent overloading the amplifier. The input signal can usually be either ac- or dc-coupled.

8. The events counter simply counts the number of pulses during a chosen time period. This counter uses the input amplifier and shaping circuits, the gate and control circuits, and the decade counting and display units.

9. Frequency is defined as events per unit of time. To measure frequency with the electronic counter, we simply count pulses (cycles) for a standard time. Ideally, this time is exactly 1 s.

10. The period meter simply counts the number of standard time pulses during one cycle of the unknown waveform. Period measurements are often used when frequency measurements will not give enough resolution. They are also used when you wish to know time rather than frequency.

11. The time-interval meter measures the time between a start signal and a stop signal. This meter has the same resolution as the period meter.

12. The period-average function reduces error found in periods measurements because of a noise on the input signal. Period-averaging requires a longer measurement interval but it produces much greater accuracies.

Chapter Review Questions

9-1. Frequency is defined as
(A) The time of one complete period regardless of wave shape (B The average time of one complete period (C) The time between two user-selected events (D) The number of complete cycles in a standard time

9-2. Period is defined as
(A) The time of one complete cycle regardless of wave shape (B) The average time of one complete cycle (C) The time between two user-selected events (D) The number of complete cycles in a standard time

9-3. Time interval is defined as
(A) The time of one complete period regardless of wave shape (B) The average time of one complete period (C) The time between two user-selected events (D) The number of complete cycles in a standard time

9-4. A special oscillator produces one pulse every 10 μs. The frequency of this oscillator is
(A) 50 kHz (B) 100 kHz (C) 500 kHz (D) 1 MHz

9-5. The earth rotates on its axis once in every 24 h. This is an example of a(n) ____?____ measurement.
(A) Frequency (B) Period (C) Time-interval (D) Event

9-6. What is the purpose of the decade counting and display unit?

9-7. In the following list ____?____ are not part of a decade counting and display unit.
(A) Latches (B) Decoders (C) Oscillators (D) BCD counters

9-8. When the total number of pulses are more than the decade counter can count, ____?____ is used.
(A) A latch (B) A decoder/driver (C) A flip-flop (D) An overrange circuit

9-9. List crystal oscillators, ovened crystal oscillators, and temperature-compensated crystal oscillators in their order of increasing temperature stability.

9-10. The purpose of the electronic counter's time-base oscillator is to give the instrument
(A) A stable voltage reference (B) A stable time reference (C) Frequency-measurement capability (D) An external reference for general laboratory use

9-11. The time-base oscillator in your counter ages a maximum of +1 ppm/year. The oscillator frequency is 10 MHz. After 1 year you would expect the oscillator frequency to be
(A) 9,999,990 Hz   (B) 9,999,999 Hz   (C) 10,000,001 Hz
(D) 10,000,010 Hz

9-12. The ovened oscillator is usually used on ____?____ counters.
(A) Portable (B) Laboratory (C) Low-cost (D) Frequency

9-13. What is the purpose of the electronic counter's divider chain?

9-14. The percentage error at the output of the divider chain is
____?____ the percentage error at the input of the divider chain.
(A) The same as (B) Less than (C) Greater than (D) A divided
percentage of

9-15. How many decades of divider are necessary to generate a 0.1-s time
reference from a 10-MHz time-base oscillator?

9-16. What is the purpose of the gate and control circuits?

9-17. The display time control is usually set so that no more than
____?____ measurement(s) can happen in 1 s.
(A) 1 (B) 2 (C) 5 (D) 10

9-18. If you want a special electronic counter which can open and close
its gate and still keep a running total, you must turn off the
____?____ pulse.
(A) Memory (B) Reset (C) Time out (D) Input

9-19. What is the purpose of the input amplifier and shaping circuit?

9-20. Hysteresis in the shaping circuit
(A) Sets the maximum signal level (B) Keeps the circuit from
triggering on noise (C) Sets the maximum frequency (D) Sets the
amplifier gain

9-21. Explain how the trigger level control and the trigger level switch
work to give triggering over all the waveform.

9-22. The events counter does not use the
(A) Decade counting and display unit (B) The time-base oscillator
(C) The input amplifier and shaping circuits (D) The gate and
control circuits

9-23. The gate of an events counter is opened by
(A) The input signal (B) The time-base chain (C) The start/stop
switch (D) The reset circuits

9-24. The gate of the frequency counter is opened by
(A) The input signal (B) The time-base chain (C) The start/stop
switch inputs (D) The reset circuits

9-25. A 7-digit 250-MHz electronic counter is used to measure the frequency of a 222.60025-MHz transmitter. It is set for a 100-ms
gating interval. What does the display read?

9-26. The gate of a period meter is opened by
(A) The input signal (B) The time-base chain (C) The start/stop
switch (D) The reset circuits

9-27. You measure the horizontal oscillator of a black-and-white TV with
a resolution of 1 μs. The counter reads
(A) 16.667 (B) 15750 (C) 63 (D) 60

9-28. The time-interval meter is just the same as the ____?____, but it
uses two inputs.
(A) Frequency counter (B) Period meter (C) Multiple-period
meter (D) Events counter

9-29. The period-average meter is used to get rid of errors in period measurements caused by
(A) Hysteresis (B) Multiple triggering (C) Noise on the input
signal (D) Too high a frequency

9-30. If a sine-wave signal has a 40-dB signal-to-noise ratio, the period-measurement error will be
(A) 3% (B) 1.0% (C) 0.3% (D) 0.1%

9-31. If the signal of question 9-30 is averaged 10 times, what will the error be?

Answers to Self Tests

1. *A*
2. *C*
3. *B*
4. *D*
5. *C*
6. *B*
7. *D*
8. *B*
9. *C*
10. *A*
11. *C*
12. *D*
13. *C*
14. *D*
15. *C*
16. *B*
17. *D*
18. *C*
19. *D*
20. *B*
21. *C*
22. *A*
23. *B*
24. *D*
25. *C*
26. *D*

31. *D*
32. *A*
33. *D*
34. *C*
35. *C*
36. *D*
37. *B*
38. *A*
39. *D*

27. The output square wave is inverted.

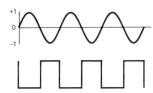

28. *C*

29. The dotted line intersects either the positive or negative peaks. Figure 9-32 shows one solution.

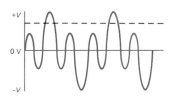

30. The ac-coupled waveform will be symmetrical at about 0 V, but the trigger point still remains at +1 V. Therefore, the output pulse becomes narrower.

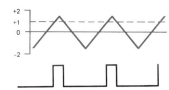

40.

Gate Interval	Resolution	Reading	Units
1 μs	1 MHz	145	MHz
10 μs	100 kHz	144.6	MHz
100 μs	10 kHz	144.60	MHz
1 ms	1 kHz	144 600	kHz
10 ms	100 Hz	144 600.0	kHz
100 ms	10 Hz	44 600.00	kHz
1 s	1 Hz	4 600 000	Hz
10 s	0.1 Hz	600 000.0	Hz

41. *F*

42. *A*

43.

Time Interval	Resolution	Reading	Units
1 μs	1 μs	0 000 000	μs
10 μs	10 μs	10 000.00	ms
100 μs	100 μs	10 000.0	ms
1 ms	1 ms	10.000	ms
10 ms	10 ms	10.00	s
100 ms	100 ms	10.0	s
1 s	1 s	10	s
10 s	10 s	0.01	ks

44. The frequency measurement takes 1000 s to resolve the required 2.0 Hz. The period measurement takes 0.5 s.

45. *B*

46. *C*

47. *B*

48. *B*

49. The period average is the fastest because 1000 periods takes 4.4 s and gives more than enough accuracy. The frequency meter needs a 100-s time base.

50. *C*

51. *B*

52. 10,000 period averages gives an accuracy of 0.00003%, but 15 ppm is 0.0015%; therefore, the time base causes the most error.

53. *D*

54. *A*

Electronic-Counter Specifications and Features

- This chapter describes the specifications, features, and errors of the electronic counter. A knowledge of these characteristics is essential in using this instrument.

  In this chapter, you will become familiar with the counter's specifications for frequency range and input sensitivity, impedance, and protection. You will also learn the specifications for time-base accuracy, gating intervals, and display time. In addition, you will become acquainted with the special features available on electronic counters as well as some common errors made by these instruments.

10-1 INTRODUCTION

The specifications and features of your electronic counter tell you what the counter is capable of doing. They tell you how accurately you are able to measure the signals you are working with. A review of the specifications for any particular electronic counter will allow you to tell if it is the right instrument to do the job which you have.

Many of the specifications do not change with the electronic counter's function; some do. Others are not important in certain modes. The electronic counter has some special errors which you will not find on other instruments. We will look at errors like the ± 1 count error to see how it happens and how to work with it.

10-2 FREQUENCY RANGE

Most of the time you will use your electronic counter in the frequency-meter function. Therefore, one of the very important specifications is the counter's frequency range. Depending on the instrument, there may be more than one frequency range for the particular electronic counter.

A simple digital frequency meter with one input has a simple specification. The manufacturer might indicate, for example, that the counter will work on signals from 10 Hz to 110 MHz. That is, the high frequency limit and the low frequency limit are specified. Any signals within this range are acceptable.

If, however, you have a more complex electronic counter, there may be as many as three inputs. Then both an upper and lower frequency may be specified for each input. The lower frequency limit disappears if the counter can be dc-coupled on one or more of these inputs.

You must look for a different frequency specification for each input, as well as looking for dc coupling on each input.

10-3 INPUT SENSITIVITY

In many ways, input sensitivity and frequency range are quite closely related. Many manufacturers specify these two on a graph such as the one shown in Fig. 10-1. Input sensitivity specifies the minimum amplitude signal which the electronic counter can measure. The signal level is given as the rms value of a sine wave. This is to say, you should expect somewhat less sensitivity when you are measuring waveforms much different from a sine wave. For example, if you are measuring a pulse train with a very low duty cycle, the sensitivity will be quite a bit less than the rms sine-wave sensitivity would lead you to believe.

Referring to Fig. 10-1, you can see that the sensitivity changes with frequency. Input sen-

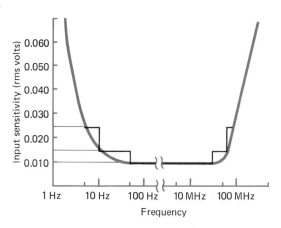

Fig. 10-1 An electronic counter's input frequency-response curve. The straight lines show the guaranteed sensitivity versus frequency. The other curve shows the typical sensitivity versus frequency for this instrument.

sitivity may be given in a number of ways. The manufacturer may say the instrument has a given sensitivity over a certain frequency range. For example, an instrument may have 15 mV sensitivity from 10 Hz to 30 MHz.

In many cases, the manufacturer will not specify exactly what happens outside this frequency range. The instrument does not "drop dead" just above its upper frequency limit or just below its lower frequency limit. It will probably quit at frequencies a few percent away from these frequency limits, however. Again, refer to the graph in Fig. 10-1. This will give you some idea of how the sensitivity changes outside the frequency limits.

Some manufacturers specify a different sensitivity for different but overlapping frequency ranges. For example, a manufacturer may specify a digital frequency meter in the following way. The instrument has 10 mV sensitivity, 50 Hz to 30 MHz; 15 mV sensitivity, 10 Hz to 60 MHz; and 25 mV sensitivity, 5 Hz to 80 MHz. The curves in Fig. 10-1 show you exactly what the manufacturer means by this more complicated specification.

If the manufacturer does not give you a set of curves, draw your own. Then you will be sure you understand exactly what is meant by the input-sensitivity specification. A simple way to look at this counter would be to say that it has 25 mV sensitivity over its full frequency range. However, this is not a safe practice. At some frequencies it has a much better sensitivity. At these frequencies, for example, the instrument could trigger on noise which is

only 10 mV. Therefore, you must know what the sensitivity is at all frequencies.

10-4 INPUT IMPEDANCE

Most counter inputs have one of two input impedances. The common input impedances are 1 MΩ and 50 Ω. The 1-MΩ input is used when you wish to avoid "loading" the circuit you are measuring. Additional 1-MΩ input specifications show the capacitance you will find in parallel across the counter's input terminals. This capacitance is usually between 10 and 50 pF. The 1-MΩ input is very handy because you may use a normal oscilloscope divide-by-10 probe on this input. When you use an oscilloscope probe, you decrease the counter's input sensitivity by a factor of 10 as well as decrease the circuit loading by a factor of 10. Normally 1-MΩ inputs are used up to about 100 MHz.

The 50-Ω inputs are for use with coaxial transmission lines. The 50-Ω input is a purely resistive input. It is generally found on inputs used for very-high-frequency applications. For example, the C input, which is usually the UHF/VHF input for most electronic counters, is often a 50-Ω nonreactive input.

The 50-Ω input will properly terminate a 50-Ω transmission line. This means that RF signals or pulse signals connected to this input are properly terminated. Therefore, no reflections will occur at the input. Reflections often cause multiple counting or what seems like a decrease in the counter's sensitivity.

10-5 INPUT PROTECTION

Normally the manufacturer will indicate the maximum rms voltage which may be applied to any particular input. This maximum voltage is much greater for 1-MΩ inputs than it is for 50-Ω inputs. Often 1-MΩ inputs with input attenuators can be connected to 100- to 400-V signals. Most 50-Ω inputs are limited to a maximum of 5 V rms. A 5-V signal on a 50-Ω input means the input must dissipate a power of 5 W. A 400-V signal on a 1-MΩ input only requires the input to dissipate a power of 0.16 W. Obviously, the 50-Ω input cannot be used as a termination or "dummy load" for a radio transmitter. It is used only to keep coaxial-cable transmission line operating properly when you are trying to make high-frequency measurements.

From page 120:
Multiple ranges

Dc coupled

Sensitivity changes with frequency

On this page:
Sensitivity changes outside frequency limits

1-MΩ input

Oscilloscope probe

50-Ω input

Terminating 50-Ω lines

Input protection

Self Test

1. Your new counter has 15 mV sensitivity from 50 Hz to 70 MHz, 25 mV sensitivity from 10 Hz to 120 MHz, and 50 mV sensitivity from 5 Hz to 160 MHz. Draw a sensitivity versus frequency curve on a graph similar to the one in Fig. 10-1.

2. Your new frequency meter has two inputs. They are called input A and input B. One of the two inputs has a 10-Hz to 110-MHz frequency range. The other input has a 50- to 500-MHz frequency range. Which input is the 50-Ω input? Why did you make this choice?

3. The input-sensitivity specification tells you the ____?____ input-signal level the counter will work at.
   A. Highest
   B. Lowest
   C. Safest
   D. Average

4. A 1-MΩ input is used because it does not load a high-impedance circuit. This resistive input is paralleled by
   A. A capacitance
   B. A resistance
   C. An inductance
   D. A nonreactive load

5. You are using a counter with a ×1, ×10, ×100 attenuator. In the ×100 position, the sensitivity is multiplied by 100. That is, there is a divide-by-100 voltage divider before the input amplifier. This counter has the input which was specified in self test question 1. The sensitivity in the ×10 position at 90 MHz is
   A. 10 mV
   B. 25 mV
   C. 50 mV
   D. 100 mV
   E. 250 mV
   F. 500 mV
   G. 1000 mV
   H. 2500 mV
   I. 5000 mV

10-6 TIME-BASE ACCURACY

As we noted in previous discussions, the accuracy of the time base primarily depends on the design of the time-base oscillator. The following list compares the accuracies which you may expect from oscillators found in electronic counters.

1. *Simple crystal oscillator:* Temperature stability—10 ppm, 10 to 40°C. Aging rate—10 ppm per year. Set tolerance—1 ppm
2. *TCXO:* Temperature stability—1 to 0.1 ppm, 10 to 40°C. Aging rate—1 ppm per year. Set tolerance—1 to 0.01 ppm
3. *Ovened oscillator:* Temperature stability—0.1 to 0.001 ppm, 0 to +55°C. Aging rate—0.1 to 0.01 ppm per year. Set tolerance—0.1 to 0.0001 ppm

As you can see, for each oscillator three specifications are given. The manufacturer tells you the error you can expect due to changes in temperature surrounding the instrument. This is called the *ambient-temperature rating.* Manufacturers also tell you how much the crystal will age, that is, how far the crystal will drift as it grows older, or its *aging rate.* Unfortunately, aging occurs most rapidly at the beginning of the crystal's operating life. It will settle down a great deal as the crystal becomes older. The manufacturer also tells you how close to the desired frequency you may easily set the oscillator. This is called the *set tolerance.*

10-7 GATING INTERVALS

The standard gating intervals which you can have on a particular counter will be indicated on its specification sheet. The gating intervals you will find on a simple digital frequency meter may be just a few. For example, you may find gating intervals as follows: 1 ms, 1 s, 10 s. However, the gating intervals available on a full electronic counter usually cover every decade from 0.1 μs to 100 s. Usually the gating-interval time is what is specified. Sometimes, the gating interval is specified by the units read. Therefore, you may find a counter with three gating intervals listed as MHz, kHz, and Hz. Such a counter is shown in Fig. 10-2.

10-8 DISPLAY TIME

Display time is the time between gate openings. It is set by the gate and control time out circuits. In some counters the display time may be adjustable. The manufacturer will then specify the adjustment range for display time. Often an adjustable display time includes an infinite position. This allows

Fig. 10-2 A low-cost electronic counter with only two gating intervals. Note that these gating intervals are marked in frequency and in time for use in the frequency meter function and the period meter function. (Courtesy of Heath Company)

you to make one measurement and then hold it until you reset the counter.

If there is no adjustable display time, the specification normally indicates only the number of readings which you can take per second. Normally this is about five readings per second.

Self Test

6. Your counter, with a TCXO time-base oscillator, has a temperature stability of +0.5 ppm/°C. What is the period of the 10-kHz output at 30°C if the instrument was calibrated at 25°C?

7. The aging rate of a crystal oscillator tells you how the oscillator frequency drifts with
   A. Temperature
   B. Time
   C. ppm/°C
   D. ppm/year

8. You can easily set your 4-MHz TCXO so the 1-MHz output is within 0.1 Hz. Therefore, the set tolerance is

A. 0.1 ppm
B. 0.025 ppm
C. 0.04 ppm
D. 0.01 ppm

9. Your counter has a 200-ms display time. You are using it on the 1-s gating interval. This means you are taking a new reading every
   A. 0.2 s
   B. 1.0 s
   C. 1.2 s
   D. 2.0 s

10. You are trying to tune an oscillator to a given frequency. Do you want the fastest or the slowest possible display time setting? Why?

10-9 FEATURES

As you can see from the earlier sections of this chapter, the digital electronic counter is a complex instrument. The electronic circuits are often further complicated by many operator-convenience features. However, many of these features are very useful. Some of them will be found only on certain instru-

123

Prescalers

Frequency ratio

Auto-ranging

Printer outputs

Battery options

Displays

ments. Some of them will be found on almost all instruments. Others will be available as extra price options. The more important of the many features you should look for are listed below. Remember, you may not be able to get all the features on one particular instrument. You should know how many you need right now and how many you can add later.

1. *Prescalers:* The prescaler may be a built-in accessory or a completely separate instrument. The prescaler divides the frequency of the signal you are measuring by some known amount. Usually this is either 4 or 10. If a prescaler is used, the decade counter and gating circuits do not have to operate at the instrument's maximum frequency. This saves money. Usually the prescaler divides by 10. All you have to do is adjust the decimal point.

2. *Frequency Ratio:* Some counters have an additional input amplifier and shaping circuit which are used in place of the time-base oscillator. When this input is used, you do not measure the ratio of your unknown frequency to the time-base frequency. You measure the ratio of your unknown frequency on one input to the unknown frequency on another input. This is a very handy input for checking digital counting circuits.

3. *Auto-ranging:* This feature is used in the frequency meter function. The gate interval is adjusted until the next to last decade counter overflows. The counter is then set on this gating interval. This is a very useful feature when you are measuring many different frequencies over a short period of time.

4. *Printer outputs:* These outputs let you connect the electronic counter to an electronic printer. This lets you record a great deal of data without writing it all down. Sometimes the printer outputs include enough connections to the counter to let you operate the counter by a computer.

5. *Battery options:* Most electronic counters operate from the power line. However, some instruments (especially frequency counters) have battery operation. Battery operation is very desirable for making away-from-the-bench measurements. If you have a battery option, you will want to know how long the counter will run on one charge of the batteries and how long it will take to recharge them.

6. *Displays:* The display information will give you the counter's digit size and type of display. Most counters use LEDs or neon displays. The instrument will also feature a number of digits. Most counters provide six, seven, eight, or nine digits. A counter with fewer than six digits means you must choose the gating intervals quite carefully to get the desired resolution. Often, you will have much more resolution than you need with a nine-digit counter.

Remember each of the features can be as important as any of the exact specifications. The features are constantly changing. For example, microprocessors are being added to laboratory electronic counters. The use of the microprocessor will make many features available on electronic counters which have not been there in the past.

Self Test

11. You know all the decade counters in one particular electronic counter are 80 MHz or less TTL. You also know the counter has a prescaler. The counter's upper frequency is specified at 250 MHz. You suspect the prescaler divides the incoming frequency by
    A. 100
    B. 40
    C. 10
    D. 4

12. You are using a six-digit auto-ranging counter in the frequency meter mode function. You are measuring two signals. One is approximately 1.2 MHz. The other is approximately 12 Hz. What two time-base gating intervals will the auto-ranging counter use? Why?

13. You are trying to do an exact study of an oscillator's aging rate. You wish to make a frequency measurement once each half hour for a week. You would like to find a counter with the _____?_____ feature.
    A. Auto-ranging
    B. Prescaling
    C. Printer-output
    D. Battery

14. You are assigned the job of setting 20 mobile radio transmitters on their exact frequency. The police department, which you are working for, does not want them removed from their automobiles. You look for an electronic counter with the _____?_____ feature.
    A. Auto-ranging
    B. Prescaling
    C. Printer-output
    D. Battery

15. You are using a low-cost counter which has five digits. However, it has a time-base oscillator which has been set to 0.1 ppm. You need to set a 10-MHz oscillator to within 1 Hz. What two gating intervals will you use to do this job? Why? How many digits of display are needed if only one gating interval is used?

10-10 ERROR IN USING THE ELECTRONIC COUNTER

The electronic counter, like many instruments, is subject to a number of measurement errors. Some errors are part of all electronic counters. Other errors come from poor operators. All the errors cause you to use the instrument wrong. This means you will not get the best possible operation from your investment.

The ± 1 Count Error

All normal digital instruments have a ± 1 count error. This error is not only found on digital counting instruments. It is also found on any instrument using a digital counting circuit. For example, we saw this error on the digital voltmeter.

What causes this ± 1 count error?

The answer is that the signal being measured is not synchronized to the time-base oscillator. This is to say, the time the gate is open is completely unrelated to the time base. The problem is identical in both the frequency and the period measurements. We will look at the problem in the frequency-measurement function. You can easily apply this information to the period-measurement function.

The reason the ± 1 count error happens is shown by the waveforms in Fig. 10-3. The waveforms in Fig. 10-3(a) and (b) show the − 1 count part of the ± 1 count error. The waveforms in this figure have exactly the same frequency. The gate intervals are also both exactly 1 s.

The only difference is the time the first zero crossing happens. Remember, the digital circuits count only one pulse at each zero crossing. In Fig. 10-3(a), 9 zero crossings fall inside the gate open time. In Fig. 10-3(b), only 10 zero crossings fall inside the gate open time. The first zero crossing just happens a little earlier than the first zero crossing in Fig. 10-3(a). Sooner or later, Fig. 10-3(b) will look exactly like Fig. 10-3(a). This happens because the frequency is not really 10 Hz but just a little bit less.

In Fig. 10-3(c) and (d) we can see how the + 1 count error happens. In these figures the actual frequency is almost 11 Hz. So most of the time it reads 11 Hz, but some of the time it reads 10 Hz.

In Fig. 10-3(a) we almost got 10 whole cycles in the gate open time. However, we did not get 10 zero crossings in the gate time. Therefore, we showed only 9 counts. In Fig. 10-3(b) we did not quite get 10 whole cycles in the gate time. However, we did get 10 zero crossings in the gate time. Therefore, we show 10 counts.

In Fig. 10-3(c) we almost got 11 whole cycles in the gate open time. However, we did not get 11 zero crossings in the gate time. Therefore, we showed 10 counts. In Fig. 10-3(d) we did not quite get 11 whole cycles in the gate time. However, we did get 11 zero crossings in the gate time. Therefore, we show 11 counts.

What does all this mean? If your counter reads 10 Hz, you may be looking at what is shown in Fig. 10-3(b) or you may be looking at what is shown in Fig. 10-3(c). That is, the counter may read 9 Hz once in a while, or it may read 11 Hz once in a while. Therefore, you only know it is 10 Hz ± 1 count.

For a reading of 10 Hz, ± 1 count is a 10% error. At a reading of 1000.000 kHz, ± 1 count is an error of 1 ppm, probably about the same error as your time base. If you have a nine-digit counter and if you read a 1-gigahertz (1-GHz) signal, the ± 1 count is only 1 part per billion.

Remember, the ± 1 count error happens on all kinds of counter measurements. It also happens when the digital counting circuits are part of another instrument like a DVM.

±1 count error

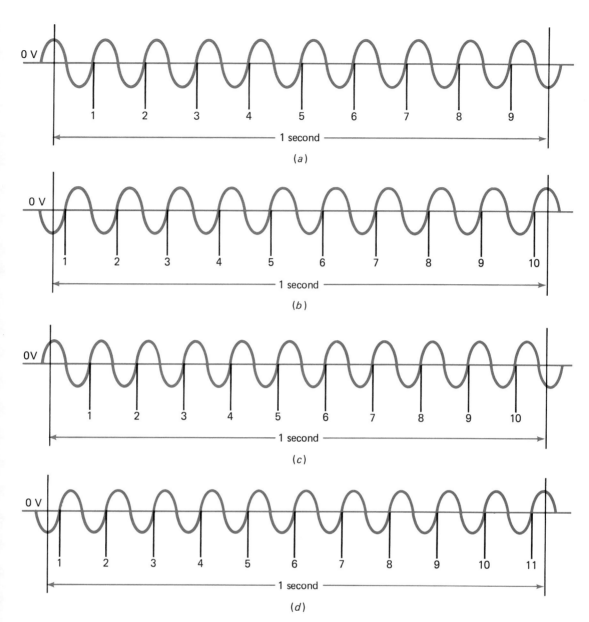

Fig. 10-3 The ±1 count error. (*a*) The electronic counter records 9 zero crossings. (*b*) The electronic counter records 10 zero crossings. (*c*) The electronic counter records 10 zero crossings again. (*d*) The electronic counter records 11 zero crossings.

Self Test

16. Your counter is in the period meter mode. It is displaying the number 1010.001. You know actually the reading could be
    A. 1010.1
    B. 1009
    C. 1010.000
    D. 1009.999

17. You are measuring a 100-MHz signal using the 1-ms gating interval. The ±1 count error is an error of

    A. 100 ppm
    B. 10 ppm
    C. 1 ppm
    D. 0.1 ppm

18. A seven-digit electronic counter is used to set a 10-MHz oscillator to within 0.1 ppm. What gating interval must be used to be sure the ±1 count error is at least no more than a 0.01 ppm error in the display? The gating interval is
    A. 100 ms   C. 10 s
    B. 1 s   D. 100 s

The Dangers of Averaging

The electronic counter can have a number of problems because of averaging. The frequency meter function is an excellent example of the errors you may have because of averaging. You can easily change this thinking to period measurement once you see the problem. Remember, the frequency meter averages the number of pulses over some period of time.

For example, a 20-Hz measurement normally means we have measured 20 evenly spaced pulses over a time period of 1 s. This is shown in Fig. 10-4. Also shown in Fig. 10-4 is a serious error which comes from making this assumption. Here we find five low-frequency pulses are followed by a burst of 10 pulses. They are at a much higher frequency. The burst is followed by five more low-frequency pulses. But the counter only sees 20 pulses over the time interval of 1 s. Therefore, it displays a frequency of 20 Hz.

By looking at this waveform, you can see that the average frequency is 20 Hz. That is, on the average there is a signal with 20 pulses per second. At any particular point in time the frequency is 20 Hz. It is as low as 5 Hz and as high as 40 Hz.

One way of detecting this condition is to change the gate time interval. For example, using a 100-ms gating interval, we might first have measured the high frequency. The next gate interval would then have happened during the low-frequency portion of the waveform. This would be displayed as a low frequency. The display, therefore, would keep changing. When the frequency randomly jumps from a low frequency to a high frequency, you may be measuring a signal whose frequency is not stable. If you think this is happening, you probably should check the waveform with an oscilloscope.

Self Test

19. When you are making a period measurement, the counter counts a series of pulses from the time-base oscillator. Why are you safe in taking this "average" measurement?

20. You are trying to measure the frequency of a 146-MHz FM transmitter. You are using a six-digit counter. For the first reading you use the 1-ms time interval. The display seems stable. For the second measurement you use the 0.1-s gating interval and the display seems unstable when you speak into the transmitter's microphone. When you use the 1-s gating interval, the display seems stable even when you speak into the transmitter's microphone. What is happening?

21. You are trying to measure the frequency of an FM sweep generator. The display on your electronic counter does not appear stable. What is the problem?

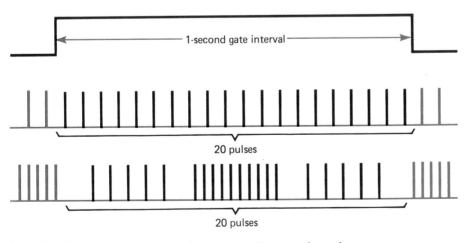

Fig. 10-4 The dangers of averaging. In the first waveform the frequency meter records 20 evenly spaced pulses. In the second waveform the frequency meter records 20 pulses at two very different frequencies. In both cases the display is still the same.

Summary

1. The manufacturer will specify the frequency range, the sensitivity, the input impedance, and the protection for the instrument's input. Often frequency range and sensitivity are tied together.

2. The time-base accuracy specification is given in three parts: temperature stability, aging rate, and set tolerance. The manufacturer will also tell you the instrument's gating intervals and display time.

3. Many special features are often available on an electronic counter, including pre-scalers, frequency ratio, auto-ranging, printer outputs, battery options, and displays.

4. The ± 1 count error happens because we do not know the last digit exactly. When you look at any one last digit, it could go either way. The ± 1 count is a large error for small display numbers. It is a very small error for large numbers.

5. Often the electronic counter assumes it is measuring a signal which is constant over time. Always remember that the counter displays an average reading.

Chapter Review Questions

10-1. The input specifications for an electronic counter do not include its
(A) Sensitivity (B) Frequency range (C) Oscillator stability
(D) Impedance

10-2. If the electronic-counter frequency range has both upper and lower limits, you know
(A) It is a 250-MHz counter (B) It has a trigger level control
(C) It has a 50-Ω input (D) It is ac-coupled

10-3. The electronic counter's input-sensitivity specification tells you
(A) The lowest-level signal which will trigger the input (B) The lowest-level signal in its frequency range which will trigger the input (C) The highest-level signal which will trigger the input
(D) The highest-level signal in its frequency range which will trigger the input

10-4. The 1-MΩ input is used on high-impedance circuits. The 50-Ω input is used
(A) To load transmitter outputs (B) To match coaxial cables
(C) On low-frequency signals (D) In dc-coupled situations

10-5. The input attenuator on an electronic counter's input is used to _____?_____ the instrument's basic input sensitivity.
(A) Decrease (B) Increase (C) Maintain (D) Stabilize

10-6. The _____?_____ is not a time-base oscillator specification.
(A) Temperature stability (B) Aging rate (C) Set tolerance
(D) Gating interval

10-7. You have a 100-MHz, five-digit electronic counter which only measures frequency. The longest gating interval is 10 s. You would expect the shortest gating interval to be
(A) 1 s (B) 100 ms (C) 10 ms (D) 1 ms

10-8. A prescaler is used to
(A) Provide period averaging (B) Provide low-cost high-frequency response (C) Divide the input signal by 10 (D) Divide the input signal by 4

10-9. An electronic counter with auto-ranging
(A) Automatically selects the correct input attenuation (B) Au-

tomatically selects the correct time-base stability (C) Automatically selects the best gating interval (D) Automatically selects the correct number of periods to average

10-10. Why does an electronic counter have a basic ± 1 count error?

10-11. The ± 1 count error becomes a ± 1 ppm error on a ____?____ digital counter.
(A) Four- (B) Five- (C) Six- (D) Seven- (E) Eight-

10-12. When the electronic counter is used in the frequency function, it can read in error because of
(A) The wrong gating interval (B) Too long a display time (C) The averaging effects of a frequency counter (D) Ac coupling

Answers to Self Tests

1.

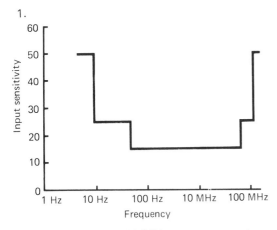

2. The 50- to 500-MHz input is the 50-Ω input. The high-frequency inputs are usually the ones where coaxial cables are used and where capacitive loading on the 1-MΩ input would be a serious problem.

3. *B*
4. *A*
5. *E*
6. 99.99975
7. *B*
8. *A*
9. *D* (*Note:* You must also let the time base recycle. This takes 1 s after the gate closes. During the 1-s time the 200-ms time out expires.)
10. Fastest. You want the display to update as quickly as possible to show the new fre-quencies as you tune the oscillator.

11. *D*
12. 10 ms and 10 s. The 10-ms interval will display the 12-MHz signal as 12000.0 kHz, thus using up the six digits. Assuming 10 s is the longest gating interval available, 12 Hz will be 0012.00.
13. *C*
14. *D*
15. To display 10 MHz to 0.1 ppm requires seven digits. Because this counter only has five digits, the 1 ms will be used to display 10 MHz as 10000. kHz. This lets us know the oscillator is at 10 MHz, not 8 MHz, 9 MHz, or some other whole number multiple of 1 MHz. The 1-s gating interval is then used to get to within 1 Hz. The display will read 00000 Hz when the oscillator is finally set to exactly 10 MHz.

16. *C*
17. *B*
18. *C*
19. Because you know the time-base oscillator is stable and noise-free.
20. The 0.1-s measurements are showing the frequency modulating produced by your voice modulating the transmitter. The deviation is probably too little for the 1-ms gating interval to show, and the 1-s gating interval averages the modulation out.
21. The FM sweep generator's output level is too low or the FM modulation is turned on.

The Cathode-Ray Oscilloscope

- This chapter discusses the cathode-ray oscilloscope. This instrument is the most important piece of test equipment in electronics.

  In this chapter, you will learn how the oscilloscope can be used to measure voltage, current, period, frequency, and time interval. You will also learn to draw a block diagram of the oscilloscope, showing the relationship between its major components and discussing their characteristics. You will also become familiar with the concepts of sweep, calibrated time base, and the dual-trace oscilloscope.

11-1 INTRODUCTION

Your oscilloscope is the most valuable tool you have to find problems in electronic equipment. It is one of the few instruments you must have to design or service electronic equipment. With the oscilloscope you can measure voltages, measure the time needed for an event to take place, measure frequency, and observe the waveforms necessary to diagnose circuit problems.

The oscilloscope presents an electronic signal as a graph. In most cases, the oscilloscope shows a picture of how voltage changes with time. If there is an electrical signal you need to know something about, just change the signal into a voltage. You can then use an oscilloscope to determine how the characteristics of the device you are measuring vary with time.

You will soon realize that the oscilloscope is a very complicated instrument. You will find it is complicated only because there are many different things you can do with it. When you thoroughly understand the blocks that make up oscilloscope circuits and what each one is supposed to do, you will be able to get the maximum use of your oscilloscope.

11-2 HISTORY

The first oscilloscopes were developed just before World War II. During the war the oscil-

loscope developed further because of the need to work on pulse circuits in radar and other military equipment. After World War II, a number of manufacturers started building oscilloscopes for the general instrumentation market. Heath introduced the first kit oscilloscope in 1947. Tektronics introduced the first of the world's largest line of oscilloscopes in 1952. Low-cost oscilloscopes did not change much during the 1950s and 1960s. They were all built with vacuum tubes.

The oscilloscope has improved a great deal in the past few years. What was an advanced oscilloscope in the mid-1960s is an old product today. For example, the recurrent-sweep oscilloscope was found in almost every television service shop in the sixties. Today, most television service shops think you must have an oscilloscope with a triggered time base. The triggered time base makes better measurements than the recurrent-sweep oscilloscope. Today it costs the same or less than the recurrent-sweep oscilloscope. A typical low-cost oscilloscope is shown in Fig. 11-1.

Most oscilloscopes are universal electronic service instruments used in repair shops and laboratories. There are also special oscilloscopes. For example, the television broadcasting industry uses special oscilloscopes to check the quality of a television picture signal. Special oscilloscopes are used in automotive repair shops. These are used to find

Fig. 11-1 A modern low-cost oscilloscope. This low-cost oscilloscope gives the service shop an instrument which was used only in the laboratory as recently as the mid-1960s. (Courtesy of Heath Company)

problems in automotive ignition systems and other parts of the electrical system. Oscilloscopes are being used to aid in medicine. These medical oscilloscopes are connected to special pickups on the patient. These collect very small electrical signals made by brain waves, heart movements, and other muscle movements. Pictures of these signals on the oscilloscope are used to monitor the functioning of various parts of the body as well as to diagnose abnormal conditions in the patient.

Self Test

1. The oscilloscope may be considered ____?____ instrument.
   A. A medical instrument
   B. An automotive instrument
   C. A special-purpose instrument
   D. A general-purpose instrument

2. The first oscilloscopes were built with all-vacuum-tube technology. Modern oscilloscopes still contain one vacuum tube. This is the cathode-ray tube on which waveforms are displayed. The rest of the modern oscilloscope is circuitry that is
   A. Passive
   B. Vacuum-tube
   C. Solid-state
   D. All of the above

3. The automotive oscilloscope is just one example of an application of the oscilloscope that is
   A. High-voltage
   B. Special-purpose
   C. General-purpose
   D. All of the above

11-3 OSCILLOSCOPE FUNDAMENTALS

Before we study the circuits which make up the oscilloscope, a few basic concepts must be understood.

The Electronic Graph

Let us examine the data in the table shown in Fig. 11-2(a). Here we see a series of voltage readings taken from an experiment. Each voltage was measured at a different point in time. Looking at these data we can see the voltage is zero at time zero. It then increases to higher and higher values through 5 time units. Between t = 5 and t = 6, it goes from +10 V to −8 V. Between t = 6 and t = 9, this negative voltage decreases to almost zero. At t = 10, the measurements of t = 0 through 9 begin to repeat.

We can get a great deal of information from reading this table. For more information we can plot a curve on a graph. The graph has voltage as the vertical axis and time as the horizontal axis. This gives a series of points like those shown in Fig. 11-21(b). The information between the points may be filled in. This makes a smooth curve like the one in Fig. 11-2(c). In many ways this smooth curve gives us a better "picture" of exactly what this voltage is doing. For example, simply observing this graph, we can see that the voltage passes through two identical cycles. Looking at this curve, we can write down a number of important characteristics about this electrical signal. We can identify the peak positive voltage, the peak negative voltage, the period of one cycle, and therefore the frequency of the waveform. If needed, we can find the length of time required for any portion of the cycle.

For example, the time length of the negative-going curve (called a *ramp*) from t = 5 is 1 time unit. The time length for the positive-going portion of the signal above 0 V is 5 time units. The signal is above 0 V from t = 0 through t = 5.

To be displayed on an oscilloscope the waveform must be cyclical. A cyclical voltage changes exactly the same way for each identical period of time. Any one cycle is the same as any other cycle. We can lay one set of measurements directly over another without changing the display.

This plot of voltage versus time is exactly

From page 130:
Low-cost oscilloscopes

Recurrent-sweep

Triggered time base

Special oscilloscopes

On this page:
Plot a curve

Cyclical waveform

Cathode-ray
tube (CRT)

Phosphor

Gun assembly

Time	Voltage	Time	Voltage	Time	Voltage
0	0	7	−6	14	8
1	+2	8	−4	15	10
2	+4	9	−2	16	−8
3	+6	10	0	17	−6
4	+8	11	2	18	−4
5	+10	12	4	19	−2
6	−8	13	6	20	0

(a)

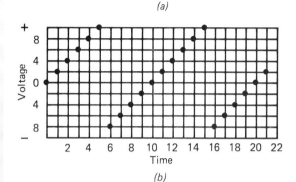

(b)

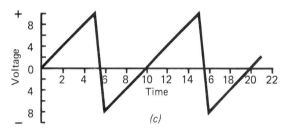

(c)

Fig. 11-2   Hand plotting a graph. (a) The data.
(b) The points on a graph. (c) A smooth curve.

what the oscilloscope is all about. One way to
do this job is to make a series of voltage measurements and then, using a piece of graph
paper, hand-plot the data. In many cases,
this is unrealistic because the waveform
happens too fast. For example, if the time for
the waveform in Fig. 11-2 was 100 ms, you
could not make one measurement every 10 ms
as the data table requires. This then is the job
of the oscilloscope. The oscilloscope displays
a graph of voltage versus time for waveforms
which repeat themselves the same way cycle
after cycle.

Self Test

4. The oscilloscope graph shows how voltage
   changes with
   A. Frequency
   B. Time
   C. Period
   D. Events

5. The oscilloscope draws one curve on top
   on another. The waveform which repeats
   itself like this is
   A. Cyclical
   B. A sine wave
   C. A square wave
   D. A graph

6. The peak-to-peak voltage in Fig. 11-2(c) is
   A. 10 V
   B. 8 V
   C. 18 V
   D. 4 V

7. The time length of the positive-going ramp
   in Fig. 11-2(c) is ____?____ time units.
   A. 9
   B. 7
   C. 5
   D. 3

8. The voltage in Fig. 11-2(c) is positive for
   ____?____ time units.
   A. 6.6
   B. 5.5
   C. 5.0
   D. 4.0

The Cathode-Ray Tube

Just as the concept of graphing a cyclical waveform is fundamental to the oscilloscope, so is
the existence of one piece of hardware. This
is the cathode-ray tube (CRT). The CRT
allows us to display the graph electronically.
The CRT is not the only way to display a
graph electronically. However, now and
probably for some time in the future, it is the
only practical way to display an electronic
graph.

   The principles of the CRT are quite simple.
As shown in Fig. 11-3, the glass faceplate of
the CRT is coated with a phosphor. Phosphor
materials emit light (glow) when they are hit by
a high-intensity electron beam. The electron
beam is generated in the *gun assembly* of the
CRT. The gun assembly is made up of filament, a cathode, a number of grids, and four
deflection plates.

   The filament and the cathode are used to
create a source of free electrons. The grids
perform two functions. First, they accelerate
the electrons using electrostatic forces. Second, they shape the electron into a thin round
beam aimed at the faceplate of the CRT.
This beam of electrons produces a small,
bright dot at the center of the CRT faceplate
as the electrons strike the phosphor coating.

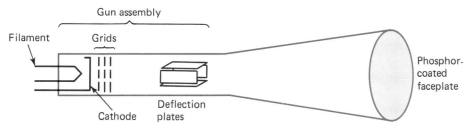

Fig. 11-3 The cathode-ray tube.

Of course, there is no information in a single bright dot. The deflection plates are used to move the dot to different places on the CRT faceplate. The CRT has two sets of deflection plates. These deflection plates also apply electrostatic forces to the electron beam. When a voltage is applied between the two sets of deflection plates, the electric field between these plates pulls the beam toward the positive plate and pushes it away from the negative one. The distance the spot moves depends on how much voltage you place between the deflection plates. If, for example, 100 V moves the spot 1 in, then 200 V will move the spot 2 in.

The *horizontal deflection plates* move the beam to the right and left of center. Therefore, the horizontal deflection plates are used to move (or *sweep*) the beam from a point well beyond the left-hand side of the tube's faceplate to a point well beyond the right-hand side of the tube's faceplate.

The other set of deflection plates are called the *vertical deflection plates*. The difference between the horizontal and vertical deflection plates is the direction the beam is deflected. The vertical plates deflect the electron beam either up from the center point or down from the center point.

We now have all the makings of an electronic graph. By applying voltages to both the horizontal and the vertical plates, the electron beam may be made to strike the CRT faceplate at any point. Figure 11-4(*a*) through (*f*) shows how the spot will be moved when different voltages are applied to the CRT deflection plates. If the voltages on the vertical and horizontal plates change, the point on the CRT faceplate where the phosphor coating glows will change.

If the voltages on the vertical and horizontal plates change fast enough, the earlier spots on the CRT will continue to glow even as the next spot is made. Thus the CRT seems to make a continuous line or *trace* on the faceplate. The shape of the trace depends on the

varying voltages on the vertical and horizontal plates. The phosphor's *persistence* is the characteristic of the CRT which makes this possible. The persistence is how long it will glow after it is struck by an electron beam. Some phosphors glow much longer than others.

For example, the phosphor used on a black-and-white television set is a P4 phosphor. The P4 phosphor emits a bright, white/blue light when struck by an electron beam. However, it has a short persistence.

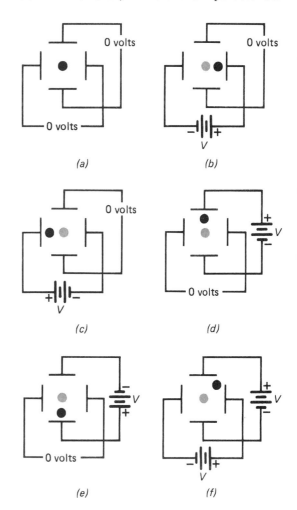

Fig. 11-4 How the horizontal and vertical deflection plates deflect the stream of electrons.

P31 phosphor

Vertical preamplifier

Vertical-deflection amplifier

Vertical position control

If it didn't, you would see "ghosts" trailing the faster-moving images on the television screen.

At the opposite end of the persistence time scale is the P7 phosphor. The P7 phosphor is a "slow" phosphor. The P7 phosphor is often used on radar displays. Frequently these are shown on television weather shows or in the movies. Here we see a slowly fading image left behind as the radar beam sweeps around the tube. This slowly fading image is a characteristic of the slow P7 phosphor.

Many different types of phosphors are in use today. One of the most common phosphors used with modern oscilloscopes is the P31 phosphor. The P31 phosphor lies between the P7 and the P4 phosphors. The persistence characteristics of the P31 phosphor used on a cathode-ray tube allow us to electronically draw a graph to present a waveform. Before the P31 phosphor was developed, the P1 phosphor was used on most oscilloscopes. The major difference between the P31 phosphor and the P1 phosphor is that the P1 phosphor is more easily burned.

Self Test

9. The P31 phosphor is a ____?____ persistence phosphor.
   A. Short-
   B. Medium-
   C. Long-

10. The vertical deflection plates move the beam
    A. Up and down
    B. From side to side
    C. Diagonally

11. The CRT uses ____?____ of deflection plates.
    A. One set
    B. Two sets
    C. Three sets
    D. Four sets

12. The electron beam moves toward the most ____?____ deflection plate.
    A. Positive
    B. Negative
    C. Neutral

11-4 THE OSCILLOSCOPE BLOCK DIAGRAM

Figure 11-5 is a simplified block diagram of a conventional oscilloscope. In the following discussion we will look at each part of this

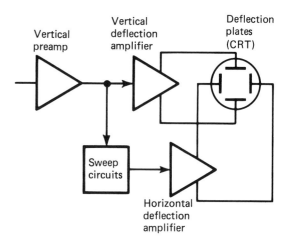

Fig. 11-5 A simplified block diagram of an oscilloscope. These basic electronic blocks plus the CRT are used to build all oscilloscopes.

block diagram. It is important to understand what it does and how it acts with the other parts of the oscilloscope.

The Vertical Amplifier

The vertical amplifiers are voltage amplifiers. They simply amplify the voltage of the incoming signal enough to deflect the electron beam up and down the face of the CRT. The vertical amplifier must have enough gain to amplify any signal to produce the deflection voltage needed by the CRT. Of course, the amplifier must have the same gain at all the frequencies at which you expect the oscilloscope to work. Often both gain and high-frequency response are hard to get, making the oscilloscope's vertical amplifier a difficult design problem.

Usually the vertical amplifier is divided into two separate parts. These are called the vertical preamplifier and the vertical-deflection amplifier.

The basic vertical oscilloscope amplifier takes positive-going input signals and deflects the beam up from the center line. Negative-going signals cause the beam to deflect below the center line.

Sometimes you do not want to use the center line as 0 V on your display. A vertical position control is added to the vertical amplifier. This control lets you set the beam position with no signal on the input. The vertical position control lets you move the beam from the center to just past the top or just past the bottom of the CRT faceplate. When measuring an input signal, you must first know

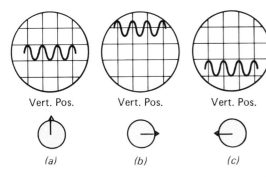

Fig. 11-6 The vertical position control. (a) **With the control at the center of its rotation the trace is in the center of the screen.** (b) **With the control turned in the clockwise direction the trace moves to the top of the CRT.** (c) **With the control turned in the counterclockwise direction the trace moves to the bottom of the CRT.**

where the vertical position control is set. Figure 11-6(a), (b), and (c) shows different settings of the vertical position controls.

As mentioned earlier, the vertical amplifier must have the same gain at all frequencies. This includes dc (frequency equals zero). For some measurements you might want to look at 50-mV peak-to-peak ripple on a 50-V dc power supply. The dc part of the signal is unimportant. The 50-mV ripple is an ac waveform of either 60 or 120 Hz.

To look at this waveform, we ac-couple the oscilloscope. That is, a capacitor is placed in series with the input signal. The ac-coupled oscilloscope does not respond to low-frequency ac signals or to dc signals.

Some low-cost oscilloscopes do not have dc-coupled vertical amplifiers. That is, they are ac-coupled all the time. The all-ac-coupled vertical amplifier is easier to design than the dc-coupled amplifier and is less expensive to build. Most oscilloscopes do not use the ac-only vertical amplifier because there is a great need to make both ac and dc measurements.

The Horizontal Amplifier

The horizontal deflection plates of the CRT are driven by the horizontal-deflection amplifier. The horizontal-deflection amplifier is just like the vertical-deflection amplifier. For some higher-frequency oscilloscopes, the frequency response of the horizontal-deflection amplifier can be less than that of the vertical-deflection amplifier.

Like the vertical amplifier, the horizontal amplifier deflects the beam depending on the input signal. Normally a positive-going signal will deflect the horizontal beam to the left of center. A negative-going input to the horizontal amplifier deflects the beam to the right of center.

The horizontal amplifier also includes position controls. The horizontal position control can move the beam from the center to either the extreme right or the extreme left side of the CRT.

The horizontal amplifier is normally used to amplify the output of the oscilloscope's sweep circuit. The sweep circuit generates a ramp or sawtooth signal which moves the beam linearly from left to right on the CRT. You can also select an external input. When you connect a signal to the external horizontal input, you can control both the X and the Y axes of the electronic graph. This is often called the XY mode. Most oscilloscopes which have an external X input have user-selected ac- or dc-coupled horizontal amplifiers. Like the vertical amplifiers, the horizontal amplifier may only be ac-coupled on very low-cost oscilloscopes.

Self Test

13. The gain of the vertical-deflection amplifier is ____?____ frequency.
    A. Proportional to
    B. Constant with
    C. Inversely proportional to

14. The horizontal amplifier amplifies the ____?____ signal.
    A. Unknown input
    B. Sweep
    C. Blanking

15. The oscilloscope time base is generated by the oscilloscope's
    A. Sweep circuits
    B. Power supply
    C. Horizontal amplifier

16. The horizontal sweep signal is a
    A. Sine wave
    B. Triangle wave
    C. Ramp (sawtooth)

Sweep Circuits

Our earlier discussion of the electronic graph showed how the horizontal axis must be calibrated in units of time. To do this, the beam starts at the extreme left-hand side of the CRT face and moves evenly to its extreme right-hand side. For example, suppose an oscilloscope faceplate has 10 vertical divisions. The time required for the beam to pass from the first dividing line to the second dividing line must be the same time as needed to move the beam from the second dividing line to the third dividing line.

The signal which moves the beam horizontally on the cathode-ray tube is called the *sweep* or *time base*. The sweep circuits or the time-base circuits are used to generate this signal. The most common form of this signal is graphed in Fig. 11-7. At the start (T_0), the signal deflects the beam all the way to the left-hand side of the CRT. The signal then increases linearly, moving the beam to the center of the CRT at T_1. The signal continues increasing linearly to a voltage which moves the beam to the right-hand side of the CRT at T_2.

It is very important to remember that the horizontal and vertical signals applied to the CRT deflection plates do not affect each other. That is, the horizontal position of the beam is the same no matter what voltage is applied to the vertical plates. Therefore, the sweep signal moves the beam uniformly from left to right across the face of the CRT. This happens no matter what the vertical signal is.

Once the beam reaches the right-hand side of the CRT, the sweep is finished. The beam must return to the left-hand side of the CRT as quickly as possible. The waveform in Fig. 11-7 shows this is done by quickly returning the voltage to the start (T_0) value. The beam is moved back to the left-hand side of the tube as fast as the horizontal amplifier and the sweep circuit can do it. Returning the beam to the left-hand side of the CRT is called *retrace*.

Often the oscilloscope has a *blanking* circuit. The blanking circuit makes sure the beam is turned off during retrace. Retrace is the time when the beam is being returned to the left-hand side of the CRT screen. Retrace time is T_2 to T_3 in Fig. 11-7. If the beam is not turned off during retrace, a faint line is drawn across the face of the CRT.

The time required for the sweep waveform to go from T_0 to T_2 is the sweep time. The sweep time is adjusted by the user. This adjustment allows you to see part of a cycle, a full cycle, or a number of cycles of the vertical waveform.

Triggering the Sweep

We can now see how an external voltage can be used to create the Y (vertical) axis of the graph. This voltage is applied to the vertical amplifier which deflects the beam above and below the centerline of the CRT. The X (horizontal) axis or time line of the graph is provided by the sweep circuits which drive the horizontal amplifiers. The beam is deflected from left to right time after time.

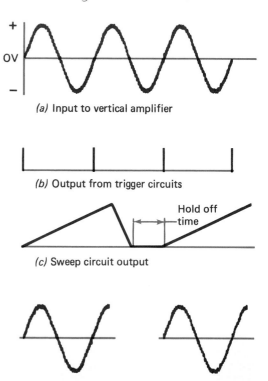

(a) Input to vertical amplifier

(b) Output from trigger circuits

(c) Sweep circuit output

(d) Osciloscope display of input waveform

Fig. 11-8 The displayed waveform. (*a*) The waveform in the vertical amplifier. (*b*) The trigger pulses. (*c*) The sweep ramp. (*d*) The displayed waveform.

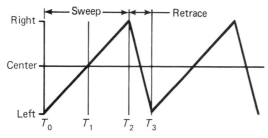

Fig. 11-7 The basic sweep signal. At time T_0 the trace is at the left of the CRT. At time T_1 the trace is at the middle of the CRT. At time T_2 the trace is at the right of the CRT. At time T_3 the trace is returned to the left of the CRT.

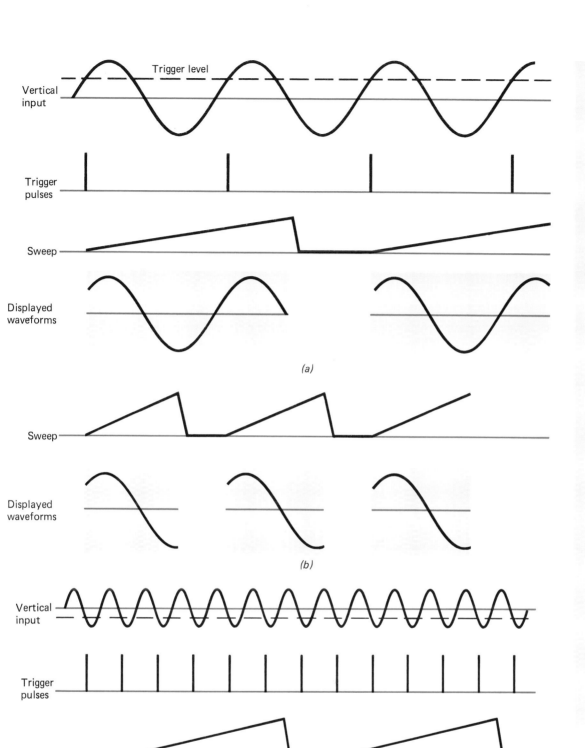

Time-base
switch

Fig. 11-9 Setting the sweep length. (*a*) Looking at more than
one cycle. (*b*) Looking at part of a cycle. (*c*) Looking at five
cycles.

Trigger circuit

Half-off time

Time-base
switch

We now have a new problem. Suppose we wish to look at a common waveform such as a 60-Hz sine wave. Each time we paint the graph on the CRT it must overlap the last one. To do this we must control the sweep start point. We must tell the time-base circuits exactly when to begin sweeping across the CRT. This job is done by the trigger circuit. The trigger circuit samples the vertical signal and drives the time-base circuits. These triggering circuits provide the synchronizing capability. That is, they make sure that the sweep across the CRT always starts at the same point on the vertical amplifier waveform.

Figure 11-8 compares the waveform connected to the vertical amplifier, the output waveform of the trigger circuits, and the output waveform of the sweep circuits and the oscilloscope's display. The trigger circuits generate a pulse each time the vertical waveform passes through a set voltage. In this case, we have chosen the point that the vertical waveform passes through 0 V. The sweep circuits are triggered by this pulse. Once triggered, the sweep starts. Until the sweep is complete, all other triggering signals are locked out. Once the sweep is done the beam is returned to the left-hand side of the tube. A small amount of time (called the hold-off time) allows the sweep circuit to stabilize. Once the hold-off time passes, the next trigger pulse starts another sweep. You can see this will start the sweep at exactly the same point on the vertical waveform as the previous one did.

Looking at the waveform shown in Fig. 11-9, you can see that the next trigger pulse may happen at many places. For example, in Fig. 11-9(*a*), the sweep time plus the hold-off time takes $1\frac{1}{2}$ cycles of the input waveform. Therefore, the sweep circuits wait for $\frac{1}{2}$ cycle of the input waveform. When the third trigger pulse is received, the sweep starts again. In Fig. 11-9(*b*) the sweep lasts only a small portion of one cycle on the incoming waveform. The sweep circuits are ready to be triggered when the next pulse occurs.

In Fig. 11-9(*c*) the user decided to look at five and one-half complete cycles of the vertical waveform. Retrace and hold-off happen during the sixth cycle of the vertical waveform. The seventh cycle of the incoming waveform restarts the sweep. In this example, only the first and seventh trigger pulses start the sweep.

The other trigger pulses are not used because the sweep is already on.

Self Test

17. The purpose of the trigger circuit is to ____?____ the sweep generator to the vertical waveform.
    A. Compare
    B. Synchronize
    C. Hold
    D. Calibrate

18. Trigger pulses which come during the time the sweep is going are
    A. Amplified
    B. Not used
    C. Synchronized
    D. Inverted

19. If the sweep lasts $4\frac{1}{4}$ vertical cycles, triggering will occur on the ____?____ trigger pulse.
    A. Third
    B. Fourth
    C. Fifth
    D. Sixth
    E. Seventh

20. Triggering will occur on every vertical cycle when the sweep period is ____?____ the vertical cycle.
    A. The same as
    B. Shorter than
    C. Longer than
    D. Synchronized with

11-5 THE CALIBRATED TIME BASE

To make the oscilloscope more useful, the time-base circuit is calibrated. That is, the time for the oscilloscope to sweep from the left-hand side of the screen to the right-hand side of the screen is exactly known. Normally the screen is divided into 10 horizontal spaces. These are shown in Fig. 11-10. The 11 vertical lines divide the faceplate into 10 horizontal divisions. A set of vertical divisions are also shown. They will be discussed later. A front-panel switch allows you to select the time for the beam to sweep from one vertical line to the next. This is called the *time-base switch*. For example, a typical time-base setting is 1 ms/div. This means the oscilloscope beam will take 10 ms (1 ms/div times 10 divisions) to travel from the left-hand side of the screen to the right-hand side of the screen.

By now you can see that the sweep genera-

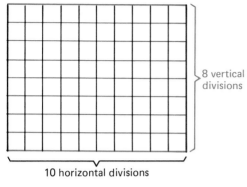

Fig. 11-10 The CRT screen. The normal CRT screen or graticule is divided into 8 vertical divisions and 10 horizontal divisions.

tor is nothing more than a special-purpose oscillator. This oscillator produces a waveform called a *sawtooth*. The reason this waveform is called a sawtooth is easily understood when you look at the diagram of a horizontal waveform. The time-base switch lets you set the period of the sawtooth.

Of course, the calibrated time base is extremely important. It allows you to make time measurements on incoming signals. For example, the oscilloscope time base in Fig. 11-11 is set to 1 ms/div. The incoming waveform repeats itself once every three divisions. We can see that the period of this waveform is 3 ms. The waveform has a period of 3 ms because it repeats itself every three divisions. Knowing that the period of the waveform is 3 ms allows you to compute the frequency of the waveform as

$$f = \frac{1}{T} = \frac{1}{0.003 \text{ s}} = 333 \text{ Hz}$$

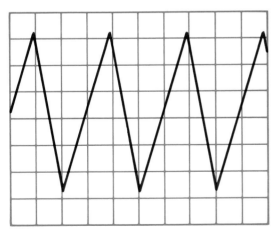

Fig. 11-11 A 333-Hz triangle wave. The oscilloscope time base is set to 1 ms/div.

The time-per-division settings on the oscilloscope's time base usually cover a wide range. This allows you to observe signals of many different frequencies.

The three displays in Fig. 11-12 show different ways of displaying a 100-kHz waveform.

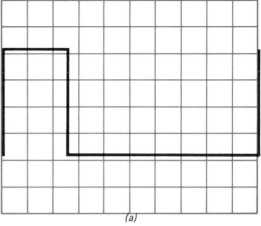

Time-per-division

Graticule

(a)

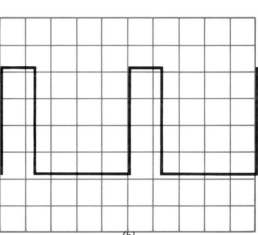

(b)

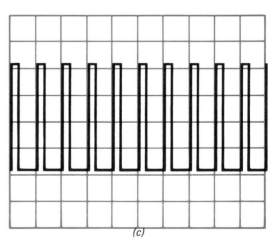

(c)

Fig. 11-12 A 100-kHz waveform. (a) The time base is set at 1 μs/div. (b) The time base is set at 2 μs/div. (c) The time base is set at 10 μs/div.

In Fig. 11-12(*a*) the time-base setting is 1μs/div. Figure 11-12(*b*) shows the oscilloscope display when the time base is set at 2 μs/div. The diagram in Fig. 11-12(*c*) shows the display for a time-base setting of 10 μs/div. You can see that these different time-base settings allow you to control the detail of the waveform you are viewing.

Self Test

21. What is the frequency of the waveform in Fig. 11-12(*c*) if the time base is set to 200 ms/div? 10 ms/div? 500 μs/div?

22. What is the width of the positive pulse in Fig. 11-12(*a*)? Assume the time base is set at 1 μs/div.

23. What is the width of the negative pulse in Fig. 11-12(*a*)? Assume the time base is set at 5 μs/div.

11-6 RECURRENT SWEEP

We have just reviewed the triggered sweep oscilloscope. That is, the sweep does not begin until it is triggered by a signal from the vertical amplifier. Once the sweep is completed, no further sweep action occurs until another triggering signal comes from the triggering circuit.

Triggering circuits like this were not common on low-cost oscilloscopes prior to the early 1970s. On these early low-cost oscilloscopes, the time-base generator was a continuously running sawtooth oscillator. It generated a waveform shown in Fig. 11-13. You can see that once the first cycle of the sawtooth is complete, a second cycle starts immediately. Once the second cycle is done, a third cycle immediately starts. The recurrent-sweep circuit does not wait for a triggering signal.

So how does the recurrent-sweep oscilloscope display a meaningful graph? If the graphs drawn by the beam do not pile on top of one another, the final graph will look like that of Fig. 11-14. One way to use the recurrent-sweep oscilloscope is to adjust the time-base oscillator frequency to be exactly

Fig. 11-13 The sawtooth. This simple waveform is generated by the recurrent-sweep oscillator.

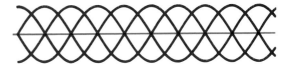

Fig. 11-14 An untriggered oscilloscope display.

the same as the vertical waveform. The waveform in Fig. 11-15 shows how this works. Such a simple system is quite limited in what it can do.

For example, in Fig. 11-9(*b*) we looked at the first 180° of a waveform using a triggered oscilloscope. This cannot be done using the recurrent sweep. There is a second way to use the recurrent sweep. The time-base frequency is set to a submultiple of the vertical frequency. This means the time base is set at one-half, or one-third, etc., the frequency of the vertical waveform.

For example, in Fig. 11-16 the sine wave displayed has a frequency of 300 Hz. The time base is set at 150 Hz so you can see two cycles on the oscilloscope screen: 300 Hz is exactly two times 150 Hz. Therefore, 150 Hz is an exact submultiple of 300 Hz. The recurrent-sweep time-base oscillator must have a continuously variable frequency. Most recurrent-sweep oscilloscopes have only a range switch and a continuously variable resistor which has frequency markings on the front panel. That is, they have only very rough calibration. The lack of a calibrated time base is a real shortcoming when using an oscilloscope. For this reason, the recurrent-sweep oscilloscope is almost gone except for the very simplest of uses.

Most recurrent-sweep oscilloscopes do have some form of synchronizing circuit. The vertical waveform is sampled and a synchro-

Input waveform

Sweep

Displayed waveforms

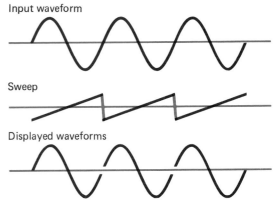

Fig. 11-15 The recurrent-sweep time-base oscillator running at exactly the vertical frequency.

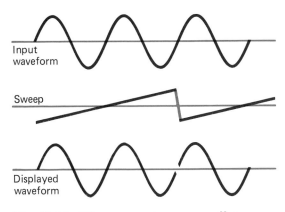

Input waveform

Sweep

Displayed waveform

Fig. 11-16 The recurrent-sweep oscillator running at one-half the vertical frequency.

nizing circuit helps keep the time-base oscillator running at the same frequency as the vertical signal or at some submultiple.

The recurrent-sweep oscilloscope cannot be set up to display $1\frac{1}{2}$ cycles of the vertical signal, for example, because you cannot synchronize the time-base oscillator at two-thirds the vertical frequency. Two-thirds is not a submultiple of the vertical frequency.

Self Test

24. A recurrent-sweep oscilloscope is displaying two cycles of a 100-Hz waveform. The frequency of the time-base oscillator is therefore set at
    A. 50 Hz
    B. 100 Hz
    C. 150 Hz
    D. 200 Hz

25. The major advantage of the recurrent-sweep oscilloscope was
    A. Vertical calibration
    B. Low cost
    C. Continuous tuning
    D. Greater accuracy

26. The recurrent-sweep oscillator is synchronized to produce
    A. The time-base calibration
    B. Error-free displays
    C. A properly synchronized blanking pulse
    D. A stable display

11-7 VERTICAL CALIBRATION

Just as the horizontal axis of our electronic graph is calibrated, so is the vertical axis. The simplified block diagram of an oscilloscope shows the vertical preamplifier and the deflec-

tion amplifier. These amplifiers are carefully calibrated on better oscilloscopes. The input to the calibrated amplifiers is usually from a calibrated vertical attenuator. Together the calibrated vertical attenuator and the calibrated vertical amplifiers allow you to make voltage measurements on the oscilloscope's vertical axis.

For example, assume the gain of the vertical amplifier is set so that a 10-mV peak-to-peak signal deflects the beam exactly one vertical division. Most oscilloscopes have eight vertical divisions (see Fig. 11-17). The signal shown in Fig. 11-17 peaks at three divisions above the center line and three divisions below the center line. We know that the signal amplitude is 10 mV/div multiplied by six divisions. This gives us a signal of 60 mV peak-to-peak.

The Vertical Attenuator

We do not want to have just one value of voltage calibration for our oscilloscope. Therefore, the calibrated vertical amplifier is combined with a calibrated input attenuator.

The vertical attenuator on an oscilloscope is just like the range switch on an electronic meter. The oscilloscope vertical amplifier is designed with all the gain needed to show the smallest signal. If the signal we wish to measure is too large, it drives the beam off the CRT. When the signal is too large, the attenuator divides the signal into a smaller value. This allows the signal to fit on the oscilloscope's screen.

For example, an oscilloscope has a vertical sensitivity of 10 mV/div. The oscilloscope has eight vertical divisions. If the signal is more than 80 mV, it will go off the screen. If you want to see a 600-mV signal, the attenuator will need to divide the input signal by 10. This

Calibrated vertical attenuator

Calibrated vertical amplifiers

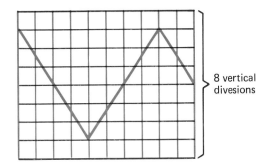

8 vertical divesions

Fig. 11-17 Vertical calibration. The vertical sensitivity is 10 mV/div.

141

Volts per
division

1−2−5 sequence

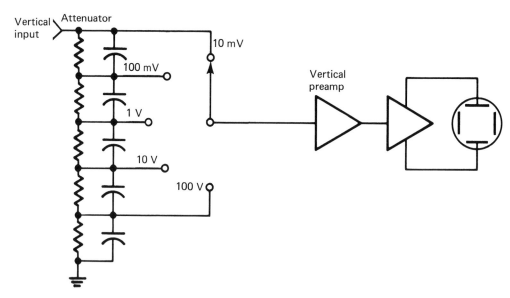

Fig. 11-18 A simple vertical attenuator connected to the vertical preamplifier. The taps represent voltage units per division of the vertical scale.

reduces the 600-mV signal to 60 mV. The 60-mV signal will fit in six divisions. If you wanted to look at a 3000-mV signal (3 V), the attenuator could be made to divide by 50. You would then have a six-division display. The 3000-mV signal divided by 50 is also 60 mV.

The input signal is connected to the input attenuator as shown in Fig. 11-18. Note that this attenuator is fully compensated just like the electronic meter's attenuator. Again the oscilloscope attenuator must be compensated to give the same attenuation over the oscilloscope's complete vertical frequency range.

The oscilloscope time base has a wide range of time which the operator can select. The vertical attenuator also has a wide range of attenuations which you can select. These allow you to use signals from a few millivolts to hundreds of volts. On the very simple oscilloscopes the vertical attenuator positions are often labeled ×1, ×10, and ×100. These indicate attenuations of 100, 10, and 1. On more expensive service and laboratory oscilloscopes the attenuator is calibrated directly in volts per division. That is, the front-panel indication of the attenuator setting tells you exactly how many volts are required to deflect the beam one vertical division on the CRT.

A common set of vertical attenuator settings is 10 mV, 20 mV, 50 mV, 100 mV, 500 mV, 1 V, 2 V, 5 V, 10 V, and 20 V/div. This is called a 1−2−5 sequence. This 1−2−5 sequence lets you see signals as low as 2 or

3 mV on the 10-mV range. On the 20 V/div range you can see signals as high as 160 V (20 V/div × 8 div).

Self Test

27. The oscilloscope's attenuator is set at 200 mV/div. What is the peak-to-peak amplitude of the signal in Fig. 11-19? What is the peak negative amplitude if the centerline is 0 V?

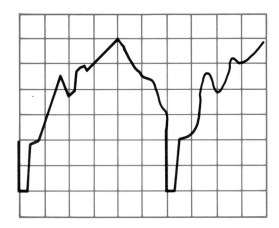

Fig. 11-19 Self test questions 27 and 29.

28. The oscilloscope's vertical attenuator is set to 50 mV div. The attenuator sequence is 1−2−5. What is the next greater sensitivity? What is the next lesser sensitivity.

29. The vertical attenuator is set at 1 V/div. What is the peak positive signal in Fig. 11-19 if 0 V is one division below the centerline?

30. The purpose of the oscilloscope's vertical attenuator is to
   A. Compensate for incorrect vertical amplifier ac gain
   B. Provide a signal tap for the trigger circuits
   C. Reduce the amplitude of signals which would overload the vertical amplifier
   D. Add extra gain to signals which cannot deflect the beam over the full CRT faceplate

31. The vertical amplifier of an oscilloscope is calibrated so that a 20-mV signal moves the beam from the center of the CRT up four divisions. Another way of saying this is to say that the oscilloscope is calibrated for
   A. 5 mV/div
   B. 10 mV/div
   C. 20 mV/div
   D. 40 mV/div

11-8 THE DUAL-TRACE OSCILLOSCOPE

Using the oscilloscopes discussed up to this point, we can make voltage measurements and timing and frequency measurements and observe the shapes of single waveforms. We cannot make time and waveform measurements between two different waveforms in two different circuits. Figure 11-20 shows a situation where this is a real problem.

The upper waveform in Fig. 11-20 is the input signal to a BCD decade counter. The lower waveform is the signal found at the fourth or *D* output of the BCD decade counter. Suppose you wish to find out which pulse is the second input pulse in the BCD decade counter. Somehow you must be able to

Input to BCD counter

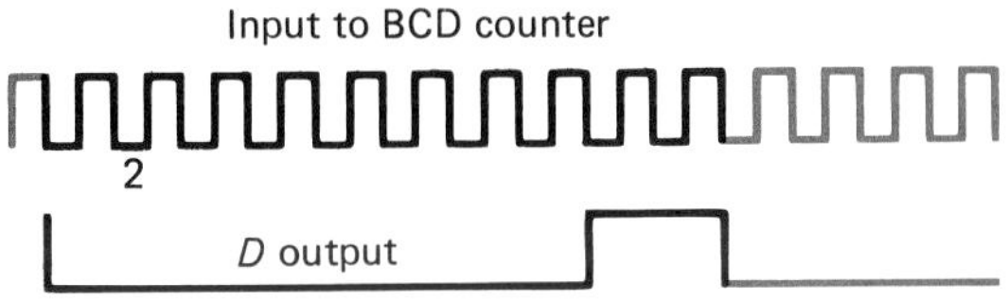

Fig. 11-20 The input and output waveforms of a BCD counter. To properly look at the timing between these pulses, you must use a dual-trace oscilloscope.

count pulses in the top waveform. Looking at the top waveform in Fig. 11-20 will not tell you the answer to this problem. The pulses shown in the bottom waveform do not show the input pulses at all; so they cannot give the answer either.

One way to solve this problem is to trigger the oscilloscope on the trailing edge of the lower waveform. The time base is then adjusted to display one or more cycles of this output waveform. The oscilloscope's input probe is then moved to show the input signal to the decade counter. Knowing that the sweep starts on the trailing edge of the output waveform, you know that the pulse on the left-hand side of the oscilloscope's display is the first pulse. You then can count to the second pulse. Needless to say, this is a difficult way to make a simple and often needed measurement.

This problem and others like it are common. They occur often in servicing or designing digital circuits. But this is not the only place where you will need to compare two different waveforms. For example, suppose you wish to look at the waveform of a signal as it goes into a filter and as it comes out of a filter. Again, you can do this with a single-trace oscilloscope. However, connecting the single-trace oscilloscope to the input waveform and then to the output waveform does not give you any information about timing changes the filter may have made on your signal. The timing between different pulses on different circuits is often extremely important in digital work.

The need to look at timing between two waveforms caused the development of the dual-trace oscilloscope. The dual-trace oscilloscope allows you to look at two waveforms exactly as shown in Fig. 11-20. This is all done on one CRT.

The dual-trace oscilloscope is different from the previously discussed single-trace oscilloscope because it has two vertical preamplifiers.

The simplified block diagram of the dual-trace oscilloscope is shown in Fig. 11-21. Two vertical preamplifiers are shown. Each connects to the vertical deflection amplifier with a switch. The switch lets one or the other, but not both, of the vertical preamplifiers drive the vertical deflection amplifier and therefore the CRT. The switching from input preamplifier *A* to input preamplifier *B* is done electronically.

143

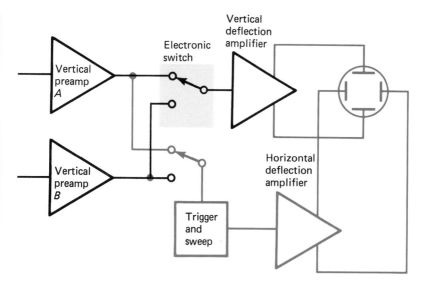

Fig. 11-21 A simplified block diagram of a dual-trace oscilloscope. Note that the outputs of the two vertical preamplifiers are connected to the vertical deflection amplifier with an electronic switch.

Self Test

32. Often you want to know if an amplifier is distorting the signal that is passing through it. One of the easiest ways to check this is to
A. Use a known good amplifier to replace the amplifier in question
B. Check the output of the suspected amplifier with a good ac voltmeter
C. Use a dual-trace oscilloscope to compare the suspected output waveform to the known good input waveform
D. Use two single-trace oscilloscopes. One is connected to the input waveform and the other is connected to the output waveform

33. Many digital circuits, such as those used in the electronic counter control circuits, must have the pulses generated in exactly the correct order. The dual-trace oscilloscope is ideal for this work because
A. It can show the exact timing relationship between two pulses in different circuits
B. Oscilloscopes are the only way to observe pulses because of their fast rise times
C. One input is for memory pulses and the other input can be used for reset pulses
D. Some electronic counters have two inputs

34. The dual-trace oscilloscope uses two vertical preamplifiers. You would also suspect that the dual-trace oscilloscope has two
A. High-voltage power supplies
B. CRTs
C. Input attenuators
D. Horizontal sweep circuits

35. Which of the following measurements does not require the use of the dual-trace oscilloscope?
A. Making relative timing measurements between two waveforms
B. Measuring the amplitude of two waveforms
C. Comparing the shapes of two waveforms
D. Telling how far apart two pulses in different circuits are

11-9 CHOPPED AND ALTERNATE MODES

There are two different ways that the dual-trace feature may be obtained. These are called the chopped mode and the alternate mode. Each method has some advantages over the other. Therefore, both are usually included on most dual-trace oscilloscopes. We will look at both of these modes individually.

The Chopped Mode

The chopped mode is most commonly used when the oscilloscope's time base is set for sweep speeds of 1 ms/div or for slower sweeps. That is to say, the chopped mode is normally used when you are looking at low-frequency waveforms.

The two signals to be compared are connected to input preamplifiers A and *B*. A high-frequency oscillator drives the electronic switch. The signal on input A is connected to the deflection plates for a short time. The signal on input *B* is then connected to the deflection plates for a short time.

The display on the CRT from this high-frequency chopping is shown in Fig. 11-22. A triangle wave is shown at the top and a sine wave at the bottom. Here the effects of chopping are greatly exaggerated. In normal conditions, you cannot see any gaps in either of the two waveforms. This is because the chopping rate is so much higher than the sweep rates. In addition, special circuits may be used to blank the beam during the time the switch is moving from one input preamplifier to the other.

Obviously, the two waveforms must be separated vertically so they don't fall on top of each other. This is done with vertical position controls in each vertical preamplifier. You will remember the vertical position controls basically add a dc offset signal to the input waveform. This means you can move either waveform up or down as you want to on the CRT screen to separate them.

The normal chopping frequency is approximately 100 kHz. At a sweep speed of 1 ms/div, this means there are 100 samples of each waveform during each sweep. This makes it almost impossible for your eye to tell the difference between 100 samples per sweep and a smooth line. If, however, the time-base sweep speed is increased to 20 μs/div, there would only be two samples per division. At this fast sweep speed the chopping becomes very noticeable. In fact, the waveform will begin to look very much like the one shown in Fig. 11-22. In order to get rid of this chopping effect on higher-frequency signals, a second form of making the dual-trace display is used for time bases greater than 1 ms/div.

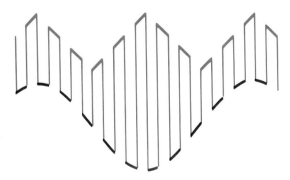

Fig. 11-22 The chopped method of presenting a dual-trace waveform. The chopping in this example is greatly exaggerated to show how it is done.

Low-frequency waveforms

Chopping rate

Vertical position controls

Chopping frequency

The Alternate Mode

The alternate mode is used when sweep speeds are greater than 1 ms/div. The switch at the output of the two vertical preamplifiers is still used. However, the switching action takes place only after one complete sweep is all done.

For example, assume you have connected two different signals to vertical preamplifiers A and *B*. Also assume you have selected triggering from input preamplifier A. In this particular example we will assume the signal connected to input amplifier A is the trailing edge of the BCD counter's *D* output. When the trailing edge of this signal occurs, the oscilloscope's time base is triggered. One complete sweep is then produced. This sweep displays the waveform shown in Fig. 11-23. When the sweep is done, the switch changes from input preamplifier A to input preamplifier *B*. The triggering, however, still is taken from input preamplifier A.

When the next trigger signal occurs, the time base sweeps again. The beam goes across the face of the tube, but this time a graph of the waveform connected to input amplifier *B* is traced on the face of the CRT. This is shown in Fig. 11-24.

As soon as this second sweep is done, the switch is again connected to the input pream-

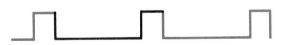

Fig. 11-23 The first sweep of the alternate-sweep mode. This is the triggering signal.

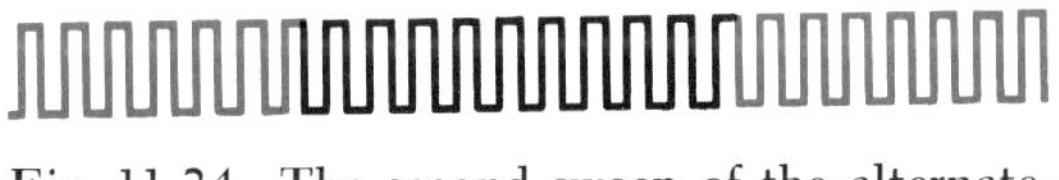

Fig. 11-24 The second sweep of the alternate-sweep mode. This sweep was triggered by the sweep shown in Fig. 11-23.

Position controls

Chopped/
alternate mode
selection

Trigger-source-
selection switch

Normal or
mixed trigger
selection

plifier A. The next (third) trigger signal causes the sweep to start again. The waveform in Fig. 11-23 is retraced on the face of the CRT.

This continues, alternating from one input to the other. Obviously, one reason this system works is the CRT's phosphor coating. The phosphor still glows for awhile even after the electron beam moves on to trace a different point on the face of the CRT. You can easily see what happens by slowing down the oscilloscope's time base. When the time base is slowed down enough, you can see the oscilloscope display one trace and then the other. Again the two traces must be vertically separated by independent position controls on each vertical preamplifier.

The Chopped/Alternate Mode Selection

Two different methods of selecting the chopped/alternate mode are common on modern oscilloscopes. Many oscilloscopes let you switch from the chopped to the alternate mode. Usually you will switch from the chopped to the alternate mode when the time per division is less than 1 ms. However, you may choose to vary from this exact time-base setting to some degree if you want to.

Other oscilloscopes automatically switch between the chopped and the alternate mode at the same specific time-base setting. Usually the 1 ms/div time-base setting is the switching point. No matter what switching point is chosen on an automatic system, the change from chopped to alternate is made automatically. You do not have any control over this operation as you did with the manual control.

Selecting the Dual-Trace Trigger Pick-Off

On the simple single-trace oscilloscope block diagram, the trigger signal is taken from the vertical amplifier. There is no question about which signal would be the trigger signal because there is only one signal to be used. The dual-trace oscilloscope allows a choice to be made.

Should you trigger from the signal in vertical preamplifier A, or should you trigger from the signal in vertical preamplifier B? Or is it best to trigger from a combination of the signals in A and B?

Because these traces are important to you when you use the dual-trace oscilloscope,

most dual-trace oscilloscopes include a trigger-source-selection switch. The most common form of the trigger-selection switch lets you trigger from the signal on channel A or trigger from the signal on channel B. This works quite well because often you do not pay close attention to which oscilloscope probe is connected to which input amplifier. It is much more convenient to switch the triggering to the desired signal than it is to interchange probes on the circuit being measured.

Some oscilloscopes offer a slightly different version of trigger pick-off. This is referred to as *normal* or *mixed* trigger selection. In one position, trigger pick-off comes from the vertical preamplifier A. In the mixed or normal mode, triggering is taken from the deflection amplifier. Of course, the deflection amplifier contains both the A and the B signals. This means that while you are observing A you are triggering on A and while you are observing B you are triggering on B.

Self Test

36. You want to look at two pulses which should be 1 μs apart. To do this you would use your dual-trace oscilloscope in the
A. Chopped mode
B. Alternate mode
C. Single-trace mode
D. High-gain mode

37. To keep the two displays on the dual-trace oscilloscope from overlapping,
A. The separate position controls are used
B. The alternate mode is used
C. The chopped mode is used
D. Mixed triggering is used

38. The two traces in a chopped-mode presentation look completely smooth because
A. The alternate mode is available for higher time-base speeds
B. The trigger pick-off signal comes from the B-input amplifier
C. The blanking circuits are used between the chopping times
D. The chopping is at a much higher frequency than the chopped waveform which you are displaying

39. Usually the dual-trace oscilloscope lets you choose the sweep-signal trigger source. ____?____ is not a typical trigger source on a dual-trace oscilloscope.

A. Channel A
B. Channel B
C. The horizontal sweep
D. Channel A and channel B mixed

11-10 OTHER MULTIPLE-TRACE OSCILLOSCOPES

The dual-trace oscilloscope is not the only multiple-trace oscilloscope. There are a number of other important multiple-trace oscilloscopes which we should briefly review.

The methods used to produce the dual-trace oscilloscope are not limited to displaying two waveforms at one time. As shown in Fig. 11-25, this method may be easily expanded to show four traces at one time. In fact, the four-trace oscilloscope has found a great deal of use in digital circuitry. There are some cases where an eight-trace oscilloscope is used.

Another method of producing the same display is to build a special CRT. This CRT has two sets of vertical deflection plates and two electron-gun assemblies. This is called a *dual-beam oscilloscope*. Two electron beams are generated and two completely separate vertical amplifiers are used to deflect these beams. Using the dual-beam design, the chopped/alternate problem no longer exists. The dual beam works the same way at low sweep speeds as it does at high sweep speeds. The major difficulty with the dual-beam oscilloscope is the cost of a dual-beam tube. The dual-beam oscilloscope usually features greater brightness than the dual-trace oscilloscope. This is because both beams are tracing the waveforms on the face of the CRT at the same time.

Four-trace oscilloscope

Dual-beam oscilloscope

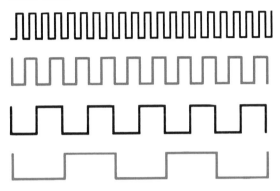

Fig. 11-25 Expanding the dual-trace mode to four traces.

Summary

1. The oscilloscope has been used as an electronic instrument for over forty years. Low-cost oscilloscope design has changed quite a bit since 1970.

2. The oscilloscope is an electronic graph. It displays voltage on the Y axis and time on the X axis. The signal displayed repeats itself exactly cycle after cycle.

3. The cathode-ray tube (CRT) is used to display the electronic graph. An electron beam generated in the gun assembly hits the CRT faceplate. The faceplate is coated with a phosphor material which glows.

4. The oscilloscope has vertical and horizontal deflection amplifiers. The vertical deflection amplifier and the vertical preamplifier amplify the input signal enough to deflect the CRT beam from the top to the bottom of the CRT faceplate. The horizontal amplifier amplifies the sweep signals so they can deflect the beam from right to left of the CRT to the left-hand side of the CRT.

5. The oscilloscope's trigger circuits make sure the sweep is synchronized with the vertical waveform. Without the trigger circuits the graph would have many different curves, not just one curve.

6. The calibrated time base allows us to make time measurements with the oscilloscope. A calibrated time base tells you the time between the divisions on the horizontal axis.

7. The recurrent-sweep oscilloscope uses a free-running sawtooth oscillator. The output of this oscillator is used as the oscilloscope's time base.

8. The calibrated vertical attenuator and the calibrated vertical amplifier together let you make a wide range of voltage measurements. The amplifier gives the oscilloscope its sensitivity. The attenuator gives the oscilloscope its range.

9. The dual-trace oscilloscope makes it possible to look at the relative timing and wave shapes of two different signals.

10. There are two ways of making a dual-trace oscilloscope presentation. Some oscilloscopes automatically select the chopped and the alternate mode. Other oscilloscopes let the operator select the dual-trace mode. Still others have a normal or mixed mode.

11-1. The oscilloscope is a general-purpose instrument. Modern oscilloscopes are all _____?_____ except for the CRT.
(A) Passive (B) Resistive (C) Vacuum-tube (D) Solid-state

11-2. For most general-purpose work the oscilloscope shows how _____?_____ changes with time.
(A) Voltage (B) Current (C) Frequency (D) Period

11-3. Except for special-purpose oscilloscopes the displayed waveform must be
(A) A sine wave (B) A graph (C) A square wave (D) Cyclical

11-4. Once the oscilloscope paints the electronic curves on the graph, you can measure
(A) Voltage (B) Frequency (C) Time (D) All of these

11-5. The CRT faceplate is covered with a phosphor material which glows when it is struck by the electron beam. Many general-purpose oscilloscopes use the _____?_____ medium-persistence phosphor.
(A) P7 (B) P31 (C) P4 (D) P16

11-6. The electron beam in a CRT is moved up and down or to the right and left by electrostatic forces applied by the
(A) Element (B) Cathode (C) Deflection plates (D) Grids

11-7. For most applications the horizontal deflection amplifier is used to amplify the _____?_____ signal.
(A) Sweep (B) Input (C) Trigger (D) Blanking

11-8. The vertical amplifier usually is a _____?_____ amplifier.
(A) Tuned (B) Narrow-band (C) Ac-coupled (D) Flat broad-band

11-9. The horizontal deflection amplifier is usually used to amplify a
(A) Sine wave (B) Ramp (sawtooth) (C) Triangle wave (D) Square wave

11-10. What is the purpose of the oscilloscope's trigger circuits?

11-11. If the oscilloscope's sweep lasts for $1\frac{1}{2}$ cycles of the vertical signal, you would expect the next sweep to be triggered by the _____?_____ trigger pulse.
(A) Second (B) Third (C) Fourth (D) Fifth

11-12. The calibrated time base lets you make _____?_____ measurements with your oscilloscope.
(A) Time (B) Pulse-width (C) Frequency (D) Voltage

11-13. Recurrent sweep was used on some older oscilloscope designs because of its
(A) Calibration (B) Low cost (C) Frequency range (D) Ease of triggering

11-14. The calibrated vertical section of an oscilloscope lets you make _____?_____ measurements.
(A) Time (B) Pulse-width (C) Frequency (D) Voltage

11-15. A common sequence for vertical attenuators and horizontal time bases is the _____?_____ sequence.
(A) 1–3–10 (B) 1–10–100 (C) 1–2–5 (D) 1.5–5–15

11-16. Why is the oscilloscope vertical attenuator compensated?

11-17. Your oscilloscope is displaying a sine wave which is almost 3 divisions peak-to-peak. The vertical attenuator is set to 100 mV/div. What is the rms value of this signal?

11-18. The dual-trace oscilloscope is often used to display cause-and-effect signals. Which of the following is not a cause-and-effect set of signals?
(A) The input/output of a decade counter (B) The input/output waveforms of a high-frequency amplifier (C) The gate and input signals on an electronic counter (D) All of the above

11-19. You would expect the increasing popularity of ____?____ is a strong cause of the increased use of dual-trace oscilloscopes.
(A) Digital circuits (B) Color television receivers (C) High-fidelity amplifiers (D) Microwave ovens

11-20. A dual-trace oscilloscope uses two different modes depending on the sweep speed used. Below 1 ms/div the chopped mode is used. Above 1 ms/div the ____?____ mode is used.
(A) High-gain (B) Wide-band (C) Alternate (D) Single-trace

11-21. The two vertical preamplifiers on a dual-trace oscilloscope have
(A) Separate position controls (B) Separate vertical attenuators (C) The same bandwidth (D) All of the above

Answers to Self Tests

1. *D*	16. *C*	26. *D*
2. *C*	17. *B*	27. 1.2 V, 600 mV
3. *B*	18. *B*	28. 20 mV/div, 100 mV/div
4. *B*	19. D Note that it started	29. 4 V
5. *A*	on the first pulse and	30. *C*
6. *C*	locked out 2–5 as the	31. *A*
7. *A*	first, second, third, and	32. *C*
8. *B*	fourth cycles were dis-	33. *A*
9. *B*	played.	34. *C*
10. *A*	20. *B*	35. *B*
11. *B*	21. 5 Hz, 100 Hz, 2 kHz	36. *B*
12. *A*	22. 2.5 μs	37. *A*
13. *B*	23. 75 μs	38. *D*
14. *B*	24. *A*	39. *C*
15. *A*	25. *B*	

Oscilloscope Specifications and Features

- This chapter describes the specifications and features of the cathode-ray oscilloscope. It is important to know these specifications and features for the same reasons that it is important to know the specifications and features of the other instruments you have studied.

  In this chapter you will learn the vertical specifications, the time-base specifications and features, and the horizontal-amplifier specifications. You will also become acquainted with basic CRT features as well as vertical delay lines, variable vertical attenuators, and variable horizontal sweep controls.

12-1 INTRODUCTION

In Chap. 11 we learned why the oscilloscope is one of the most valuable electronic instruments. We also learned that the oscilloscope is quite a complicated instrument. The more an oscilloscope can do, the more complicated it is. The oscilloscope can measure voltage, time, and frequency. We can also observe the shape of a waveform with this instrument. This much more than we can do with any of the other instruments that we have discussed in earlier chapters. It follows that any instrument which can do all these things will have many more specifications and features than the other instruments.

The oscilloscope's specifications and features are different in some ways from the specifications and features of other instruments you have studied. It is not as accurate as many of these instruments. This does not mean the oscilloscope's specifications are not important. In fact, you probably need to know more about an oscilloscope's specifications to use it correctly.

Looking back at some of the other instruments we have studied, we remember many instruments which have a lot of features. The oscilloscope is no exception to this rule. Most common oscilloscopes have quite a few features. Very good oscilloscopes have more. Depending on the job, these features can be very important or not matter at all.

12-2 VERTICAL-AMPLIFIER SPECIFICATIONS

The specifications for the vertical amplifier are the most important specifications for the oscilloscope. In fact, the vertical specifications are so important that, in most cases, the other oscilloscope specifications and features depend on the vertical-amplifier specifications and features. In this section we will look at each of the important vertical-amplifier specifications. As we review these specifications you will see that some of them are very closely tied to others. That is, in some cases if you know one specification you will automatically know the other.

Vertical-Amplifier Bandwidth

Your oscilloscope's vertical-amplifier bandwidth is the instrument's single most important specification. Usually when any one single specification is given to describe an oscilloscope, it is the oscilloscope vertical-amplifier bandwidth. For example, you might hear someone say, "A 5-MHz oscilloscope is really good enough for television servicing." The 5 MHz refers to the oscilloscope's vertical-amplifier bandwidth. In other words, the person who is speaking has used the vertical-amplifier bandwidth to describe the complete oscilloscope

The oscilloscope's vertical-amplifier-bandwidth specification tells you the maximum-frequency waveform you can expect to measure on this instrument. Note that we say this is the maximum *reasonable* frequency that the oscilloscope can be used to measure. The vertical-amplifier bandwidth is not the highest frequency at which the oscilloscope will work. That is, the vertical-amplifier-bandwidth specification is not a "drop dead" specification.

The oscilloscope vertical-amplifier-bandwidth specification tells you the upper 3-dB frequency for this amplifier. Theoretically, the oscilloscope vertical amplifier will have exactly the same voltage gain from a very low frequency to some high frequency. As the oscilloscope's vertical amplifier is used at higher and higher frequencies, it slowly begins to lose gain. At some frequency the gain will be 0.707 of the gain at the lower frequencies. That is, a signal of the same amplitude will deflect the signal only 0.707 as far as it is deflected at the lower reference frequency. The frequency at which this happens is called the vertical-amplifier bandwidth. The diagram in Fig. 12-1 shows what happens.

You can see that the vertical-amplifier gain is the same from dc to nearly 10 MHz. But as we increase the frequency, some gain is lost. In fact, the gain is 3 dB less at a frequency near 10 MHz.

Thus, the vertical-bandwidth number does not tell you everything. First, the gain of the vertical amplifier starts to decrease before the −3-dB point is actually reached. Second, there is still a lot of vertical-amplifier gain left at frequencies above the −3-dB point. Theoretically, the vertical-amplifier gain decreases at a rate of 6 dB/octave for frequencies beyond the 3-dB point. Again, Fig. 12-1 shows what this means.

For example, in Fig. 12-1 we are working with an oscilloscope which has a 10-MHz vertical bandwidth. This specification means that the vertical-amplifier gain at 10 MHz is only 0.707 of its 1-kHz value.

Looking at this curve, we can see that the oscilloscope's vertical amplifier is definitely not dead above 10 MHz. At 20 MHz (twice 10 MHz) the oscilloscope's vertical amplifier still has some gain. In fact, the gain is 9 dB less than it was at the low-frequency reference point. This is because the gain was 3 dB down at 10 MHz. The additional 6-dB loss is because the frequency increased one octave (two times). This 9-dB drop in gain means that the vertical amplifier has only one-third the gain that it did at the lower frequency. However, the oscilloscope is not completely useless at these higher frequencies, its use is simply more limited.

The oscilloscope vertical-amplifier-bandwidth specification tells nothing about what happens at low frequencies. The low-frequency response is not specified because it depends on the setting of the ac/dc coupling switch. If the oscilloscope's vertical amplifier is ac-coupled, it will not pass dc. An ac-coupled vertical-amplifier frequency response is shown in Fig. 12-2.

In this figure, we can see that the oscilloscope loses gain at both high and low frequencies. The low-frequency loss is because of the ac coupling. Obviously, the oscilloscope also has a low-frequency −3-dB point. This −3-dB point will lie between 1 and 10 Hz. If the oscilloscope has ac/dc vertical coupling, we can change between the frequency-response curves in Figs. 12-1 and 12-2 by simply flipping the ac/dc switch.

The curves in Figs. 12-1 and 12-2 show a

From page 150:
Vertical-amplifier
bandwidth

On this page:
Upper 3-dB
frequency

6 dB/octave

Low-frequency
response

Low-frequency
−3-dB point

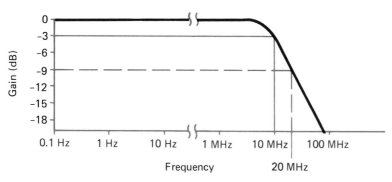

Fig. 12-1 A vertical-amplifier frequency-response curve. Note that this amplifier has "flat" gain from dc to nearly 10 MHz. Its gain is down 3 dB (−3 dB) at 20 MHz.

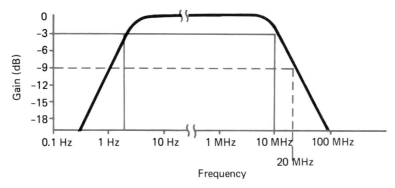

Fig. 12-2 A vertical-amplifier frequency-response curve showing the low-frequency drop (called roll-off) because of ac coupling.

theoretical vertical-amplifier response curve. Real oscilloscope vertical amplifiers do not have exactly the same gain for all the middle frequencies. There will be increases and decreases in the vertical-amplifier gain. These variations are usually not more than 1 dB (± 0.5 dB). But, remember 1 dB is 10%!

This tells you something else about the oscilloscope. It should not be thought of as a precision instrument. If you want to make a voltage measurement within a few percent, do not use an oscilloscope. It just is not that accurate.

Of course, as the frequency of your measurement gets near the oscilloscope's -3-dB frequency point, the vertical amplifier starts to lose gain. This means you are losing measurement accuracy.

If the oscilloscope vertical-amplifier-bandwidth specification is not an indication of the oscilloscope's accuracy, what does it tell you? This specification tells you what frequency waveforms you can look at and be assured that the frequency response of the vertical amplifier is not distorting the waveform. Most of the time when you use an oscilloscope your greatest interest is the waveform's shape. You will not be as interested in absolute voltage accuracy.

Self Test

1. A typical oscilloscope used to service small digital computers and other equipment using TTL typically has a 35-MHz vertical bandwidth. This means the oscilloscope's vertical gain at 35 MHz will be ____?____

times the oscilloscope's vertical gain at 1 kHz.
A. 1.414   C. 0.5
B. 0.707   D. 0.333

2. The perfect oscilloscope's vertical amplifier ____?____ over most of its useful frequency range.
A. Is within 1 dB
B. Is down 3 dB
C. Is flat
D. Is down 6 dB

3. If you are using a 20-MHz oscilloscope and you need to look at the shape of a waveform which is mostly made up of 25-MHz signals, you would expect the displayed waveform
A. To be down 4.5 dB
B. To be slightly smaller than it would be if you used a 25-MHz oscilloscope
C. To be slightly larger than it would be with a 25-MHz oscilloscope
D. To be down 6 dB from a 1-kHz reference

4. You are working in a shop which has a 15-MHz dual-trace oscilloscope. You are fixing a CB transmitter, and you think that the AM modulator may not be working. Is there any point in using the oscilloscope to look at the CB transmitter carrier to see if there is any modulation? Why?

5. A 5-V peak-to-peak 100-Hz waveform covers five vertical divisions on a 10-MHz oscilloscope. You would expect a 5-V peak-to-peak 10-MHz waveform to cover ____?____ vertical divisions on this same oscilloscope.
A. 10
B. 5
C. 3.5
D. 1.5

Vertical-Amplifier Rise Time

The oscilloscope's rise-time specification tells you how much time the vertical amplifier takes to go from 10 to 90% of a vertical change. The simplest way to show the vertical-amplifier rise time is to connect the oscilloscope to a pulse with infinitely fast rise time, that is, a pulse which goes from zero to its full amplitude instantly. There are no pulses with infinitely fast rise time, but pulses can be generated which are so much faster than the oscilloscope's rise time that you can think of them as being infinitely fast.

Figure 12-3 shows an oscilloscope display of a short pulse. This diagram shows the rise time of the pulse is the time needed to go from the 10 to the 90% amplitude points on the pulse. The sweep circuit is set to 5 ns/div.

In this example you can see that the pulse has a 10-ns rise time. It has a 5-ns fall time. Fall time is just the opposite of rise time. That is, it is the time for the pulse to go from the 90 to the 10% amplitude points. If the pulse coming into the oscilloscope has an infinitely fast rise time, we can assume that this display shows only the oscilloscope's 10 ns rise time. That is, the pulse rise time is zero; so the oscilloscope's rise time is 10 ns.

The oscilloscope vertical-amplifier -3-dB bandwidth specification and the vertical-amplifier rise-time specification are very closely tied together. In fact, if the oscilloscope vertical amplifier is built correctly, you can calculate one specification if you know the other.

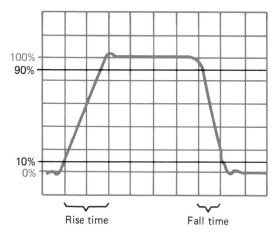

Fig. 12-3 The oscilloscope amplifier's rise time. This narrow pulse with a very fast rise time shows the vertical amplifier's rise time.

The oscilloscope vertical-amplifier frequency response and the oscilloscope vertical-amplifier rise time are related by

$$T = \frac{350}{f}$$

where T is the rise time in nanoseconds and f is the -3-dB frequency in MHz. Using this formula, we can easily calculate an oscilloscope's rise time if we know its -3-dB frequency. Also, of course, we can calculate the oscilloscope's -3-dB frequency if we know its rise time.

The following table gives some -3-dB frequencies commonly found on oscilloscopes. The rise times we expect to find on these oscilloscopes are also shown.

Frequency	Rise time
1 MHz	350 ns
5 MHz	70 ns
10 MHz	35 ns
15 MHz	24 ns
20 MHz	18 ns
35 MHz	10 ns
50 MHz	7 ns
100 MHz	3.5 ns

This relationship between the oscilloscope vertical-amplifier frequency response and the oscilloscope vertical-amplifier rise time is true only if the gain falls off at 6 dB/octave for frequencies beyond the -3-dB point. If the oscilloscope frequency response and rise time have this relationship, the oscilloscope will faithfully display the shape of any waveform which has frequencies up to the -3-dB frequency. If an oscilloscope's vertical amplifier does not have this relationship between its frequency response and rise time, some waveforms, especially pulses, will be distorted.

What do you do if the manufacturer does not specify the oscilloscope's rise time? If you are working with a relatively expensive oscilloscope, you are safe in assuming that it meets this specification. If you are working with one of the very low-cost oscilloscopes, there is a good possibility that the oscilloscope will not faithfully reproduce the exact shape of waveforms whose frequency is near the -3-dB point.

Vertical-amplifier rise time

10 to 90%

Fall time

$T = \dfrac{350}{f}$

Volts per
division

Bandwidth
versus
sensitivity

Oscilloscope
probe

Self Test

6. You are using a new 10-MHz oscilloscope to make measurements on some TTL digital circuits. The rise and fall times of most TTL circuits are usually 7 ns. You would expect to see the rise and fall times on this oscilloscope of
   A. 70 ns
   B. 35 ns
   C. 7 ns
   D. 3.5 ns

7. You have a 20-MHz oscilloscope and a fast pulse generator. You measure the oscilloscope rise time as 35 ns. When you measure the 100-μs rise-time pulse which is coming through a very low-frequency amplifier, would you expect the pulse waveform to be faithfully reproduced? Why?

8. When you begin to work with Schottky TTL, you need a new oscilloscope to replace the 35-MHz model you have been using for some time. You get a new 75-MHz oscilloscope. When the oscilloscope is delivered, you measure the rise and find it is exactly what it should be. It is
   A. 21 ns
   B. 26 ns
   C. 5 ns
   D. 2 ns

9. The oscilloscope rise-time specification tells you
   A. That the rise time is equal to 0.35 divided by the frequency
   B. How the oscilloscope will display high-frequency waveforms
   C. That a 100-MHz oscilloscope will have a 3.5-ns rise time
   D. That the oscilloscope vertical response is within 1 dB.

Vertical-Amplifier Sensitivity

The oscilloscope vertical-amplifier sensitivity specification is much like the sensitivity specification for an analog voltmeter. This specification tells how much signal is needed at the oscilloscope's input to deflect the beam a certain distance on the face of the CRT.

When we worked with the voltmeter, we used a specification which told how much voltage was needed to deflect the pointer to the instrument's full-scale position. The oscilloscope's vertical sensitivity specification is a little different. Oscilloscope sensitivity specifications indicate the voltage needed to deflect the beam one vertical division on the CRT. Remember, the face of the CRT is divided into vertical and horizontal divisions, usually 8 by 10.

Typically the oscilloscope sensitivity is between 1 and 20 mV/div. Vertical sensitivities better than 10 mV/div are found only on higher-cost oscilloscopes. Once you know what the oscilloscope's vertical sensitivity is, you can figure out how many volts are needed to deflect the beam over all the CRT's vertical divisions. All you have to do is multiply the vertical sensitivity by the number of vertical divisions. This is eight divisions for most oscilloscopes.

For example, suppose you are using an oscilloscope which has a vertical sensitivity of 10 mV/div. If this oscilloscope has eight vertical divisions, 80 mV will deflect the beam from the bottom division of the CRT to the top division of the CRT.

Sometimes you will find that the very high-sensitivity positions of an oscilloscope have a limited bandwidth. For example, an oscilloscope might have a 25-MHz bandwidth for vertical-amplifier sensitivities of 10 mV/div or less. But for vertical-amplifier sensitivities of greater than 10 mV/div, the bandwidth might be limited to 5 MHz.

The oscilloscope vertical sensitivity is a very important specification. It tells you how small a signal you can look at. Usually you can just barely begin to see signals which are 20% of the sensitivity specification. For example, if an oscilloscope has a 20 mV/div sensitivity, you just barely begin to see a 4-mV signal.

Having an oscilloscope with good vertical sensitivity is important for another reason. Often you will use your oscilloscope with an oscilloscope probe. The oscilloscope probe is used to raise the oscilloscope's input impedance. This probe reduces loading on the circuit you are measuring. Again, you do not get something for nothing. These probes usually divide the signal level by 10. Therefore, if you are using an oscilloscope with a sensitivity of 10 mV/div, a ÷10 oscilloscope probe reduces the instrument's overall sensitivity to 100 mV/div.

Like the voltmeter, the vertical-sensitivity specification tells you the most sensitivity you can expect from the oscilloscope. Sometimes you do not want all this sensitivity. The instrument's sensitivity may be reduced by using a vertical input attenuator.

The Vertical Input Attenuator

An oscilloscope vertical input attenuator is much like the input attenuator on a good analog voltmeter. It is just a simple voltage divider. Because the oscilloscope is used at much higher frequencies than most analog voltmeters, very careful compensation is needed. The input sensitivity and the range of the vertical attenuator indicate the minimum and the maximum signals your oscilloscope can be used with.

The most common vertical attenuator sequence is the 1–2–5 sequence. Occasionally you will find a low-cost oscilloscope which has only a 1–10–100–1000 attenuator sequence. Obviously, there are some difficulties with using this limited attenuator. The vertical attenuator may be split in two sections. This is shown in Fig. 12-4. Here we see the first input attenuator divides the signal by 1, 10, 100, or 1000. The second low-impedance uncompensated attenuator divides the signal by 1, 2, or 5. Using these different attenuator combinations, we may divide the input signal by 1 (that is, no division at all) to 5000 (maximum division).

Figure 12-5 shows all the different vertical sensitivity settings which we can get using these two attenuators and an oscilloscope with a 10 mV/div basic sensitivity.

The maximum sensitivity (10 mV/div) has no input attenuation. The minimum sensitivity has the full attenuation of 500. This gives us an oscilloscope with a 50 V/div sensitivity.

Suppose this oscilloscope has eight vertical divisions. When the vertical attenuator switch is in this position, it will need 400 V to deflect the beam past all eight vertical divisions. Although the vertical attenuator is built in two sections, the oscilloscope will have only one switch with 12 positions, as shown in Fig. 12-5.

Typically the vertical attenuator has an accuracy of ±3%. You must also add any other errors which may occur because of the attenuator's high-frequency response. Usually the attenuator's high-frequency errors and the lack of the vertical-amplifier gain flatness are all part of the 1-dB specification. Often the 1-dB specification is not mentioned at all.

In addition to the switch positions, oscilloscope attenuators often include a continu-

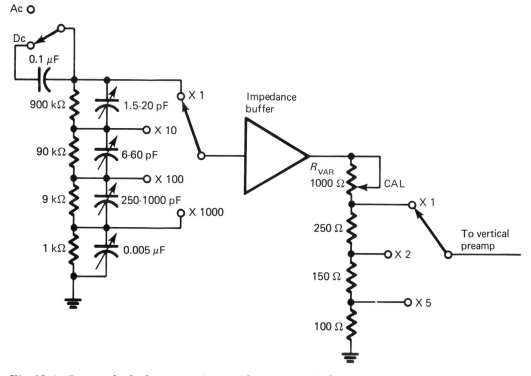

Fig. 12-4 One method of constructing a wide-range vertical attenuator. The first section of this attenuator is a high-impedance voltage divider. After the input buffering amplifier a low-impedance voltage divider is used.

155

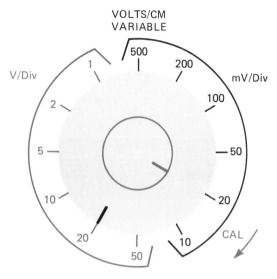

VOLTS/CM
VARIABLE

Fig. 12-5 A typical oscilloscope vertical attenuator. Turning the vertical attenuator switch clockwise increases the oscilloscope's sensitivity. The concentric center knob is the continuously variable attenuator.

Continuously
variable
attenuator

Relative
measurements

Calibrated
position

1-MΩ input
impedance

Input
capacitance

Time-base
specifications

Sweep-speed

ously adjustable variable attenuator. This is shown in Fig. 12-4 as R_{VAR}. It lets you set the attenuation at any point between the two settings you are working with.

For example, suppose your oscilloscope is set on the 2 V/div position. Using the continuously variable vertical attenuator lets you set the oscilloscope's vertical sensitivity anywhere between 2 and 5 V/div. This continuously variable vertical attenuator is useful when you are making relative measurements where absolute accuracy is not at all important. Usually the continuously variable vertical attenuator has a calibrated position. The input attenuator's accuracy specification is good only when the continuously variable attenuator is in the calibrated position.

As you can see by looking at Fig. 12-4, the oscilloscope's vertical attenuator is a compensated attenuator. This is because it must work at very high frequencies.

The input impedance of most vertical attenuators is usually 1 MΩ. The 1-MΩ input impedance has been chosen because many different oscilloscope accessories, such as probes, are designed to work into this standard impedance. One megohm also does not load most circuits. The 1-MΩ resistive input impedance is shunted by 20 to 50 pF. This capacitance comes from the compensating capacitors.

Self Test

10. You have an oscilloscope which has a basic sensitivity of 20 mV/div. The input attenuator has 12 positions in a 1–2–5 sequence. List the complete set of this instrument's calibrated vertical sensitivities.

11. This oscilloscope's vertical attenuator is set at 5 V/div. The next most sensitive position is
    A. 2 V/div
    B. 5 V/div
    C. 10 V/div
    D. 20 V/div

12. A typical oscilloscope's input impedance is 1 MΩ shunted by 30 pF. Instead of using an oscilloscope probe, you use a 3-ft length of coaxial cable with alligator clips to connect to the circuit. The cable has a capacitance of 30 pF/ft. The real input impedance is therefore
    A. 1 MΩ
    B. 1 MΩ paralleled by 30 pF
    C. 1 MΩ paralleled by 120 pF
    D. 1 MΩ paralleled by 90 pF

13. At 4 MHz the oscilloscope from self test question 12 connected to a circuit by the test cable presents a load of 1 MΩ resistive paralleled by
    A. 330 Ω capacitive
    B. 1333 Ω capacitive
    C. 440 Ω capacitive
    D. 1,000,000 Ω capacitive

14. You have an oscilloscope with a 5 mV/div sensitivity, a 1–2–5 sequence, and a 6 × 10 division CRT grid. It takes a ____?____ signal to just fill the screen when the vertical attenuator is in the third most sensitive position.
    A. 500-mV
    B. 200-mV
    C. 100-mV
    D. 300-mV
    E. 120-mV
    F. 60-mV

12-3 TIME-BASE SPECIFICATIONS AND FEATURES

The oscilloscope's horizontal *time-base* or *sweep-speed* specifications are usually quite simple. The manufacturer tells you the oscilloscope's range of sweep speeds and the accuracy of the sweep time. In addition to the sweep's range and accuracy, you also need to know what triggering controls are available.

As we indicated earlier in this chapter, often some other instrument specifications depend on the vertical specifications. The horizontal time-base specifications usually depend on the vertical specifications. For example, the fastest horizontal sweep speeds will usually change with different vertical-bandwidth specifications.

The Horizontal Time Base

The horizontal time base normally has a 1–2–5 sequence just like the vertical input attenuator. For most oscilloscopes the slowest horizontal sweep speed is between 1 s and 100 ms/div. The fastest horizontal sweep speeds are usually between 1 μs and 50 ns/div. As noted before, this range usually depends on the vertical bandwidth.

For example, an oscilloscope with a 5-MHz bandwidth often has a maximum horizontal sweep speed of only 1 μs/div. On the other hand, an oscilloscope with a 35-MHz vertical bandwidth may well have a 100 or even 50 ns/div maximum horizontal sweep speed.

While the fastest horizontal sweep speed usually depends on the oscilloscope's vertical bandwidth, the slowest sweep speed often depends on the number of horizontal sweep-speed switch positions the manufacturer uses. Typically oscilloscopes have 18, 22, or 24 sweep-speed positions. The more expensive the oscilloscope, the more sweep-speed positions there are. Often this is called the oscilloscope's *time-base* switch.

Figure 12-6 shows a 22-position horizontal time-base switch for a triggered-sweep oscilloscope. Here you can see the 21 different sweep speeds which might be found on a typical oscilloscope. This time-base switch lets you set sweep speeds from 0.1 μs/div to 500 ms/div.

There is also a switch position marked EXT (for external). When the horizontal time-base switch is set to the external position, the sweep generator is disconnected. The external horizontal input is then connected to the horizontal deflection amplifier. In Fig. 12-6 you can see that the 1–2–5 sequence is used.

The time-base markings are for the sweep speed per division. You must multiply this by the total number of horizontal divisions to find the sweep time required to cross the entire CRT face. For a typical 6 × 10 or 8 × 10

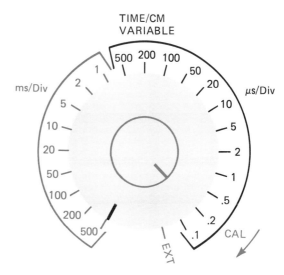

TIME/CM VARIABLE

Fig. 12-6 A 22-position horizontal time-base switch. There are 21 time-base settings and an external input position on this switch.

1–2–5 sequence

Time-base switch positions

External position

Continuously variable control

Time-base accuracy

division CRT graticule (grid), the multiplier would be 10.

Like the vertical attenuator, the horizontal time base often has a continuously variable control. That is, when this control is used, you can set the time per division to any value between two time-base settings on the time-base switch. Again, this is a very valuable feature when you are trying to make relative measurements. The horizontal time-base accuracy specifications are good only when this control is in the calibrated position.

The time-base accuracy is usually between ±3 and ±5%. The more accurate time bases, as you would expect, come with the more expensive oscilloscopes.

The Recurrent Sweep

The oscilloscope time bases we have been discussing so far are for triggered-sweep operation. If you are using one of the older oscilloscopes with recurrent sweep, you will find completely different time-base markings and controls. Figure 12-7 shows a typical set of recurrent-sweep time-base controls.

In Fig. 12-7 you can see there are two controls. The switch on the right selects the recurrent-sweep oscillator's frequency range. For example, the sweep oscillator can be continuously varied between 1 and 10 kHz if the switch is set to the third position.

The continuously variable control on the left is simply marked MIN and MAX. If the control is at the minimum position, the time-base

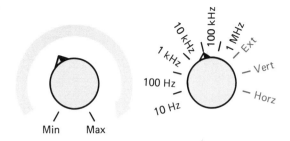

Fig. 12-7   Recurrent-sweep time-base controls. The switch settings are marked in frequency, not in time.

Trigger mode switch

Normal triggering

Automatic triggering

oscillator runs at approximately 1 kHz. If the control is at the maximum position, the time-base oscillator operates at approximately 10 kHz. Although the control can be set anywhere between these two positions, there is usually little or no calibration.

Self Test

15. You are using your oscilloscope to look at a 1-MHz square wave. You want to look at two complete cycles of the waveform. You set the time-base switch to
    A. 2 μs/div
    B. 1 μs/div
    C. 0.5 μs/div
    D. 0.2 μs/div
    E. 0.1 μs/div
    F. 0.05 μs/div

16. The fastest sweep speed on an oscilloscope is 1 μs/div. The time-base switch has 18 positions in a 1–2–5 sequence. List all 18 sweep speeds.

17. In the oscilloscope in self-test question 16, the beam will travel across the 8 × 10 division CRT graticule in ____?____ when the time-base switch is set to the lowest position.
    A. 2 s   E. 100 ms
    B. 1 s   F. 50 ms
    C. 500 ms   G. 20 ms
    D. 200 ms   H. 10 ms

18. You are using your oscilloscope in the 2 ms/div time-base position. It has a ±3% time-base accuracy. One cycle of the square wave you are looking at takes up exactly eight horizontal divisions. The period of this signal is
    A. 2 ms ± 40 μs
    B. 16 ms ± 480 μs
    C. 20 ms ± 400 μs
    D. 40 ms ± 800 μs

Triggering Controls

Figure 12-8 shows a typical set of time-base triggering controls. In this diagram we see two controls. On the right-hand side is a six-position switch. Using this switch you can select either positive or negative triggering. If you select positive triggering, the signal which starts the horizontal sweep will be taken from the leading edge of the signal in the vertical amplifier. This is shown in Fig. 12-9(*a*).

If you select negative triggering, the trigger signal will be taken from the trailing edge of the signal in the vertical amplifier. This is shown in Fig. 12-9(*b*).

Using this switch, you can also select auto, normal, or line triggering. In the normal position one trigger pulse is generated each time the waveform in the vertical amplifier passes through the trigger point on the slope you select. If there is no waveform in the vertical amplifier, there will be no trigger pulse. If there is no trigger pulse, the CRT's trace will disappear because no sweep is generated to cause a trace.

Being without a trace when there is no vertical signal often makes it difficult to operate the oscilloscope. The automatic mode of triggering is provided to help offset this. When you select the automatic (auto) triggering mode, a trigger pulse is produced every time the vertical waveform passes through the triggering point. This is just like normal-mode triggering. However, the automatic mode offers one other feature. Shortly after there is no signal in the vertical amplifier, triggering pulses are automatically generated. This keeps a trace on the CRT. This means if you lose triggering, a trace will appear shortly on the CRT.

The other setting of the trigger-mode switch is called *line*. When you select the line posi-

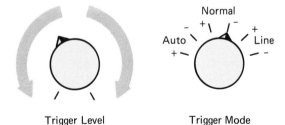

Fig. 12-8   A typical set of triggering controls. The control on the left is the trigger-level control. The control on the right is the trigger-mode control.

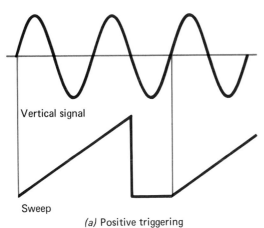

Vertical signal

Sweep

(a) Positive triggering

Vertical Signal

Sweep

(b) Negative triggering

Fig. 12-9 Positive or negative triggering. *(a)* A sweep signal positively triggered from the vertical signal. *(b)* A sweep negatively triggered from the vertical signal.

tion, trigger pulses are taken from either the leading (+) or trailing (−) edge of the 60-Hz power line. This triggers the time base in synchronization with the power line. The line position is very useful if you think that the signal you are looking at may be related to the power-line frequency. All you have to do is select either positive or negative line triggering. If this synchronizes the waveform, you know that it is related to the power-line frequency.

For example, this is an excellent way to look for ripple on power-supply lines, because ripple is directly related to the power-line frequency.

The left-hand control in Fig. 12-8 is called the *trigger-level control*. The trigger-level control lets you trigger over any 180° portion of the input waveform. The trigger-level control is used with the plus and minus settings of the trigger-mode control to trigger over any 360° portion of the vertical waveform.

On some oscilloscopes the trigger-level control is not used in the automatic mode. On these oscilloscopes selecting the automatic mode means that you have selected zero-crossing triggering.

An important triggering specification is triggering bandwidth. Triggering bandwidth specifies the smallest signal that will trigger the oscilloscope at a given frequency. On good oscilloscopes the triggering-bandwidth specification usually specifies a signal which is $1\frac{1}{2}$ to 2 times the vertical −3-dB point and has a vertical amplitude of 0.4 to 0.2 division.

For example, suppose you are using an oscilloscope with a 15-MHz vertical bandwidth and an 8 × 10 division graticule. By experimenting you find you can still trigger the oscilloscope (that is, make the trace stable) at 45 MHz when the trace is 0.2 division high. The triggering bandwidth for this oscilloscope is 45 MHz for a 0.2-division display.

You can see that this bandwidth specification is different from the specification for a vertical bandwidth. That is, if you increase the frequency above 45 MHz or lower the displayed amplitude below 0.2 division, triggering of the sweep circuits will stop. Therefore, the triggering-bandwidth specification is a "drop dead" specification.

Self Test

19. The trigger-mode switch does not let you select
    A. Trigger level
    B. Trigger slope
    C. The normal mode
    D. The automatic mode

20. The triggering-bandwidth specification tells you
    A. When the trigger signal is at the −3-dB point
    B. When the trigger signal is on the −6-dB/octave slope
    C. That the triggering will quit for a higher frequency or a lower amplitude
    D. That you will have to use a faster time-base setting

21. The trigger-mode switch line position lets you trigger the sweep circuits from
    A. A 60-Hz signal
    B. A 50-Hz signal
    C. The ac power-line frequency
    D. From either the leading or the trailing edge of the signal in the vertical amplifier

External
horizontal
input

XY mode

Horizontal
bandwidth

Round CRT

Rectangular
CRT

22. Using the trigger-level control, you can continuously adjust the oscilloscope triggering point over 180° of the vertical waveform from the negative peak to the positive peak. The ____?____ control lets you select which slope this will be, thus giving you a full 360° of triggering.
  A. Time-base
  B. Trigger-mode
  C. Continuously variable vertical attenuator
  D. Continuously variable horizontal time base

23. The purpose of the automatic mode is to
  A. Automatically select the correct triggering slope
  B. Synchronize the triggering of the sweep generator with the 60-Hz ac power line
  C. Supply trigger pulses when there is no vertical waveform to generate a sweep
  D. Eliminate the trigger-level control by automatically selecting zero crossing as the trigger point

12-4 HORIZONTAL-AMPLIFIER SPECIFICATIONS

Generally speaking the horizontal-amplifier specifications are not too important on most oscilloscopes. Most of the time you will use the internal horizontal sweep circuits as a signal source for the horizontal amplifiers. Therefore, you do not really care what the horizontal-amplifier specifications are because they are matched to the horizontal time base. Once in a while you will want to use the horizontal amplifier. At this time you will need to know its specifications.

Sometimes there will be two different sources for the horizontal-amplifier input. The simplest input is a single jack marked External Horizontal Input. This may have a × 1, × 10, × 100 attenuator. Usually, the horizontal input attenuator has a continuously variable control to let you make relative measurements.

In some oscilloscopes you will find an XY mode. Usually this feature is found on dual-trace oscilloscopes. When you select the XY mode, one of the vertical input attenuators and preamplifiers is connected to the horizontal deflection amplifier. You then have two full range input attenuator/amplifiers. One is on the Y (vertical) axis of the electronic graph and the other is connected to the X (horizontal) axis of the electronic graph.

Just like the vertical amplifier, the horizontal amplifier has a bandwidth specification. However, in most cases the horizontal-amplifier bandwidth is much less than the vertical-amplifier bandwidth. For example, horizontal bandwidths of 1 to 3 MHz are very common on oscilloscopes which have 15- to 35-MHz vertical bandwidths. This means that you can only put fairly low-frequency signals on the horizontal inputs. This also means that XY operation can take place only over a fairly limited frequency range.

12-5 OSCILLOSCOPE FEATURES

As we have noted before, the oscilloscope usually has a large number of features. It is these features which make the oscilloscope so versatile and therefore so valuable. We have already reviewed the major vertical features in discussions on the basic oscilloscope design. These are the single-trace and the dual-trace modes of operation.

The single-channel oscilloscope is very simple. The major dual-trace features are chopped and alternate operation, that is, the selection of how the oscilloscope looks at the two channels.

The CRT

One of the features which makes the oscilloscope much easier to use is the quality of the CRT. Usually the quality of the CRT is directly dependent upon the price of the oscilloscope. However, certain CRT features make an oscilloscope easier to use.

Two different types of CRTs are the round CRT and the rectangular CRT. The original and lowest-cost CRT is shaped very much like a wine bottle. The electron-gun assembly is in the narrow portion of the glass bottle. The faceplate where the phosphor is deposited is round. A round-faced CRT has the disadvantage of having its corners cut off. More modern oscilloscopes use a rectangular CRT. The glass "bottle" of the CRT is shaped to mate with a rectangular faceplate. The rectangular faceplate means that the entire oscilloscope viewing area is usable.

One of the CRT's important specifications is its brightness. The oscilloscope manufacturer does not tell you the trace brightness in units of light intensity. Instead, the voltage used to accelerate the electron beam is specified. Most simple CRTs use 2000 to 3000 V to accelerate the electron beam. These simple CRTs are called *monoaccelerator CRTs* because there is only one high-voltage electrode in the tube.

The very high-brightness CRTs have a second high-voltage electrode. It is called a postaccelerator. The postaccelerated tube is much brighter than the monoaccelerated tube. The postaccelerated tube has two points of acceleration. The first point is in the electron-gun assembly. The second point of acceleration is after the deflection plates. CRTs using a postaccelerator use acceleration voltages between 5000 and 20,000 V.

Brightness is very important when you are looking at high-frequency waveforms or if you are looking at pulses which do not occur very frequently. That is, if you are looking at low-duty-cycle waveforms, you are hitting the phosphor with more energy less frequently to make it glow. If you don't use a high-energy beam, you will not see the waveform.

Two other features will be found on some CRTs. The flat-face CRT uses a faceplate which is truly flat. Older CRTs have a slight bow in the faceplate to keep focus over the entire faceplate. This bowed faceplate introduces some distortion in the shape of the waveform. Although the flat-faced CRT is more expensive to build, it does not introduce distortion in the waveform.

In all the waveform diagrams we have shown a graticule on the CRT faceplate. The normal graticule has 8 vertical divisions and 10 horizontal divisions. Usually the graticule is clear plastic with painted or etched lines. It is just placed in the CRT faceplate. On very good CRTs, the graticule is actually etched into the inside of the CRT faceplate. This internal graticule eliminates the parallax error which you get from using an external graticule.

Many oscilloscope CRTs use an 8 × 10 graticule. Again, this means the graticule has 8 vertical divisions and 10 horizontal divisions. Once in a while you will find an oscilloscope with a 6 × 10 division graticule. Usually this is on the earlier solid-state oscilloscope designs.

Often the graticule divisions are 1 cm. Many oscilloscopes with 1-cm divisions will be marked with volts per centimeter or time per centimeter. However, not all oscilloscopes use centimeters. Some large-screen laboratory oscilloscopes use 1.2 cm/div. Some small portable oscilloscopes use 0.8 cm/div. Therefore, you should speak of divisions to be safe.

Most oscilloscopes which feature a graticule also have scale illumination. Scale illumination is simply a small lamp used to light up the graticule lines. This allows you to see the lines, giving you better visibility. Good scale illumination does not wash out the trace on the face of the CRT.

Time-Base Magnifiers

Often an oscilloscope will have a *magnifier*. The magnifier is simply an extra gain position on the horizontal amplifier. It is used during horizontal sweep. For example, you are using the oscilloscope with the time base in the 100 ms/div position. Suppose you wish to look at the waveform shown in Fig. 12-10(a) in detail. This is quite difficult because using a faster sweep-speed setting will not allow you to see all of the waveform. In this case a magnifier is used.

The ×5 magnifier gives you a 20 μs/div sweep speed. The way this is done can be seen in Fig. 12-10(b). By using the horizontal

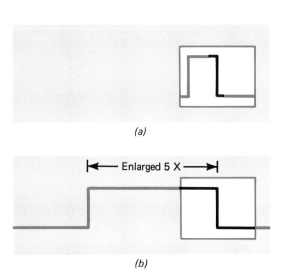

Fig. 12-10 The time-base magnifier. (a) The oscilloscope presentation without magnification. (b) The effective oscilloscope presentation with magnification. You must use the horizontal position control to look at all of the trace. You are looking at only one-fifth of the trace at any horizontal-position control setting.

161

Magnifier gains

Accuracy

Pretriggered
waveform

position control you can put any portion of the magnified trace on the screen. Horizontal magnifiers come in gains of ×2, ×5, or ×10.

If you use the magnifier, the trace becomes dimmer. This is because in the ×5 position, for example, the trace appears on the screen only one-fifth as long as it would be without the magnifier. Magnifiers also produce a trace that is less accurate than an unmagnified trace.

Vertical Delay Lines

Vertical delay lines are usually featured on medium- to high-priced oscilloscopes. The vertical delay lines get rid of a triggering-circuit problem. The problem is the time needed for a trigger circuit to work. For example, in Fig. 12-11 you can see there is a

problem in looking at the leading edge of a pulse. The leading edge of the pulse must send a triggering signal to the sweep circuits. The sweep circuits must then get started. By this time the leading edge of the pulse is gone.

This is solved using the vertical delay lines. The vertical lines delay the vertical waveform as it goes to the vertical deflection amplifier and then to the vertical deflection plates. This means that after the triggering pulse occurs, the sweep circuits have time to start before the signal that is being displayed is shown on the face of the CRT.

Most vertical delay lines allow you to see a few nanoseconds of the *pretriggered* waveform. That is, the vertical delay is more than the time needed for the trigger pulse and the sweep circuits to start. Vertical delay lines are almost a must on oscilloscopes which are for use with digital circuits. Without the ver-

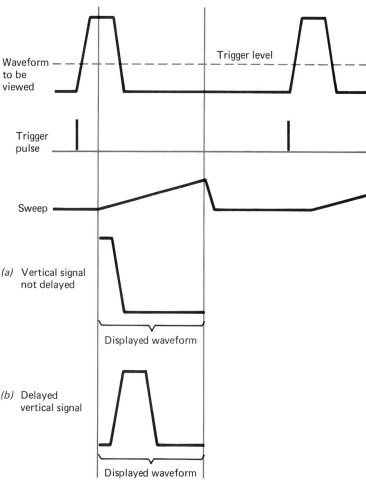

Fig. 12-11 The need for vertical delay lines. (a) The trigger pulse starts the sweep too late to display the leading edge of the pulse. (b) The delayed vertical waveform lets the sweep start before the pulse reaches the vertical deflection plates.

tical delay lines it is very difficult to do careful pulse timing.

The Package

Oscilloscope packaging is also a very important feature. An oscilloscope that is to be used in the field, should be quite portable. Usually such portable oscilloscopes have an easy carrying handle, and they may also store accessories such as probes and the line cord. Once again we do not get something for nothing. The portable oscilloscope usually has a crowded front panel and a smaller CRT.

If a portable oscilloscope is not essential, a laboratory or bench package is preferred. The laboratory or bench oscilloscope usually has more space on the front panel, a larger CRT, and often many more small convenience features than the portable oscilloscopes do. A battery-operated oscilloscope is available, but it is usually quite expensive.

Self Test

24. You are using an oscilloscope with a 15-MHz vertical amplifier. The horizontal amplifier has a 1-MHz bandwidth. You would assume the maximum usable XY bandwidth for this oscilloscope is
A. 100 kHz

B. 1 MHz
C. 5 MHz
D. 15 MHz

25. You wish to look at a 1-μs pulse which happens 10 times per second. You have two oscilloscopes to choose from. Oscilloscope A has a 15-MHz bandwidth and oscilloscope B has a 5-MHz bandwidth. Oscilloscope A uses a 3-kV acceleration voltage and oscilloscope B uses a 10-kV postaccelerator. Which oscilloscope, A or B, would you choose? Why?

26. You are using your oscilloscope in the 20 μs/div position. To get a good look at the waveform, you turn on your ×2 magnifier. The sweep speed is now
A. 10 μs/div
B. 20 μs/div
C. 40 μs/div
D. 200 μs/div

27. You are using a 50-MHz oscilloscope with vertical delay lines. The delay lines will show 50 ns of pretriggered waveform. You set the oscilloscope's triggering circuits to trigger on the leading edge of a 200-ns-wide pulse at the 10% point. The oscilloscope will display
A. The last 10% of the leading edge and the rest of the pulse
B. The pulse from the 10% point on
C. The trailing edge of the pulse
D. All the pulse plus 50 ns of the waveform before the +10% point

---

Summary

1. An oscilloscope has a vertical-amplifier frequency-response specification. This is the frequency at which the amplifier gain is 3 dB less than a low-frequency reference point.

2. The most important characteristic of the oscilloscope is how faithfully it reproduces a waveform, not how accurately it measures a voltage.

3. The oscilloscope vertical-amplifier rise-time specification is an indication of how faithfully the oscilloscope will display high-frequency waveforms.

4. Most oscilloscopes have a calibrated vertical amplifier. The vertical-amplifier sensitivity specification indicates the voltage needed to deflect the CRT's trace one vertical division.

5. Most vertical input attenuators have a 1–2–5 sequence. Often a continuously variable vertical attenuator lets you set the oscilloscope's sensitivity anywhere between two input attenuator settings.

6. The horizontal time-base specifications involve the sweep generator and the triggering circuits. First, the ranges and accuracy of the time-base sweep generator are specified. Second, the type of triggering controls and the triggering bandwidth are specified.

7. The trigger-mode switch lets you select positive or negative slope triggering in the normal, automatic, or line modes.

8. The horizontal specifications indicate the performance of the horizontal amplifier when an external signal is applied to it.

9. The CRT can be described by its shape, brightness, acceleration voltage, and faceplate. The CRT's graticule, or grid, usually has 8 vertical and 10 horizontal divisions.

10. The horizontal magnifier lets you spread the trace horizontlly. If you use a × 10 magnifier, you make the trace ten times as long as the face of the tube. The trace also is only one-tenth as bright.

11. Vertical delay lines are used to delay the signal going to the CRT by enough time to let the sweep start.

Chapter Review Questions

12-1. The oscilloscope's vertical-bandwidth specification indicates the frequency at which the gain is _____?_____ below the gain at a low-frequency reference point.
(A) −1 dB (B) −3 dB (C) −6 dB (D) −9 dB

12-2. You are using a new high-performance 50-MHz oscilloscope. The gain of the vertical amplifier at 10 MHz is _____?_____ the gain at 1 kHz.
(A) The same as (B) 2 times (C) 0.5 times (D) 0.707 times

12-3. As you check the oscilloscope's gain above its vertical-bandwidth frequency you expect the gain to roll off at a rate of
(A) −1 dB/octave (B) −3 dB/octave (C) −6 dB/octave
(D) −9 dB/octave

12-4. Give the formula that relates the oscilloscope's vertical-amplifier bandwidth and its vertical-amplifier rise time.

12-5. You have just purchased a dual-trace 25-MHz oscilloscope. What rise time would you expect in each channel? Would you expect both channels to use exactly 25 MHz? Why?

12-6. What is the purpose of the vertical-amplifier rise-time specification?

12-7. You have just purchased an oscilloscope with a very good sensitivity of 1 mV/div. Its minimum sensitivity is 20 V/div. How many positions are there on the attenuator switch?

12-8. A typical oscilloscope input impedance is
(A) 1 MΩ paralleled by 300 pF (B) 1 MΩ paralleled by 30 pF
(C) 10 MΩ paralled by 30 pF (D) 10 MΩ paralleled by 3 pF

12-9. If you can see a signal which is 0.2 division high, what is the usable range of the 8 × 10 division oscilloscope in question 12-7?

12-10. Your 50-MHz oscilloscope has a fast sweep speed of 50 ns/div and a slow sweep speed of 2 s/div. How many time-base positions does this oscilloscope have?

12-11. You want to use your dual-trace oscilloscope to look at both the input and the output waveforms of a decade counter. The decade counter is counting a 10-kHz square wave. You set the time base on
(A) 1 ms/div (B) 500 μs/div (C) 200 μs/div (D) 100 μs/div

12-12. Typically the oscilloscope's vertical attenuator and its time base have about the same accuracy. Often this is
(A) ±10% (B) ±3% (C) ±1% (D) ±0.3%

12-13. You want your oscilloscope to trigger on the leading edge of the waveform you are observing. Also you do not want to let the screen go without a trace when the signal stops for a short while. To do this you select _____?_____ triggering.
(A) +Automatic (B) −Automatic (C) +Normal (D) −Normal

12-14. You have just purchased a good 20-MHz dual-trace oscilloscope. You would expect the trigger bandwidth to be
(A) 5 MHz (B) 10 MHz (C) 20 MHz (D) 40 MHz

12-15. An oscilloscope which has very high brightness probably has a _____?_____ CRT.
(A) Round (B) Rectangular (C) Postaccelerated (D) Flat-faced

12-16. You are using an oscilloscope which has a 10-kV acceleration voltage and a flat-faced rectangular CRT with 1.2 cm/div. You would expect this is not a _____?_____ oscilloscope.
(A) Low-cost (B) 50-MHz (C) High-intensity (D) High-cost

12-17. What is the purpose of a vertical delay line?

12-18. You are using an oscilloscope with a $\times 10$ magnifier to look at a 1-MHz waveform. The time base is set on 10 μs/div. The waveform shows one complete cycle every
(A) 10 divisions (B) 5 divisions (C) 2 divisions (D) 1 division

Answers to Self Tests

1. *B*
2. *C*
3. *B*
4. Yes. Because a 15-MHz oscilloscope will be down only 9 dB at 30 MHz. That is, when you are just looking for wave shape (modulation) and have plenty of signal (a transmitter output), you will be able to look at the waveform's essential characteristics.
5. *C*
6. *B*
7. Yes. Even though the bandwidth/rise time relationship is wrong, the rise time of the pulse is so far below the upper end of the oscilloscope that only a 350-kHz os-

cilloscope is needed to observe this pulse.
8. *C*
9. *B*
10. 20 mV/div, 50 mV/div, 100 mV/div, 200 mV/div, 500 mV/div, 1 V/div, 2 V/div, 5 V/div, 10 V/div, 20 V/div, 50 V/div, 100 V/div
11. *A*
12. *C*
13. *A*
14. *E*
15. *D*
16. 200 ms/div, 100 ms/div, 50 ms/div, 20 ms/div, 10 ms/div, 5 ms/div, 2 ms/div, 1 ms/div, 500 μs/div, 200 μs/div, 100 μs/div, 50 μs/div, 20 μs/div, 10 μs/div,

5 μs/div, 2 μs/div, 1 μs/div, external
17. *A*
18. *B*
19. *A*
20. *C*
21. *C*
22. *B*
23. *C*
24. *B*
25. *B,* because the oscilloscope with the 10-kV postaccelerator will be much brighter. The pulse has a duty cycle of 0.001%, which will be hard to see. The 5-MHz bandwidth is more than adequate to see a 1-μs pulse.
26. *A*
27. *D*

Index